Hunting for Dinosaurs

The MIT Press, Cambridge, Massachusetts, and London, England

Hunting for Dinosaurs Zofia Kielan-Jaworowska

Translated from the Polish

English translation by Israel Translation Society
Copyright © 1969 by Zofia Kielan-Jaworowska

Originally published in Poland under the title "Polowanie na Dinozaury"

Set in Linofilm Baskerville.
Printed and bound in the United States of America
by The Maple Press Company, York, Pennsylvania.

All rights reserved. No part of this book may be reproduced in any form or by
any means, electronic or mechanical, including photocopying, recording, or by
any information storage and retrieval system, without permission in writing from
the publisher.

SBN 262 11030 X (hardcover)
SBN 262 61007 8 (paperback)
Library of Congress catalog card number: 73-87288

Contents

1. Introduction
   1
2. Dinosaurs and Mammals
   9
3. A Bit of History
   23
4. The First Reconnaissance Expedition
   and Subsequent Preparations
   29
5. Through the Gobi by Field Car
   35
6. The Dinosaurs of Tsagan Khushu
   49
7. Reconnaissance Trips to Altan Ula and Nemegt
   63
8. The Trans-Altaian Gobi
   73
9. Getting Ready for the 1965 Expedition
   87
10. The Altan Ula Camp
    95
11. Our Biggest Skeleton
    109
12. The Cretaceous Mammals of Bayn Dzak
    123
13. There Ain't no Such Animal
    137
14. Nemegt, 1965
    149
15. The Rhinoceroses of Altan Teli
    165

Acknowledgments

I wish to express my gratitude to Professor Roman Kozłowski, Professor Kazimierz Kowalski, and to Dr. Ryszard Gradziński having read the manuscript and for the valuable advice. Thanks are also due to Professor Kazimierz Kowalski and to Dr. Halszka Osmólska for their kind permission to publish excerpts from their reports. The majority of the illustrations were supplied by R. Gradziński and W. Skarżyński. Other photographs were taken by W. Maczek, J. Małecki, J. Lefeld, M. Czarocka, and H. Kubiak. The drawings throughout the book are by Maciej Kuczynski.

Zofia Kielan-Jaworowska
Warsaw, Poland

Hunting for Dinosaurs

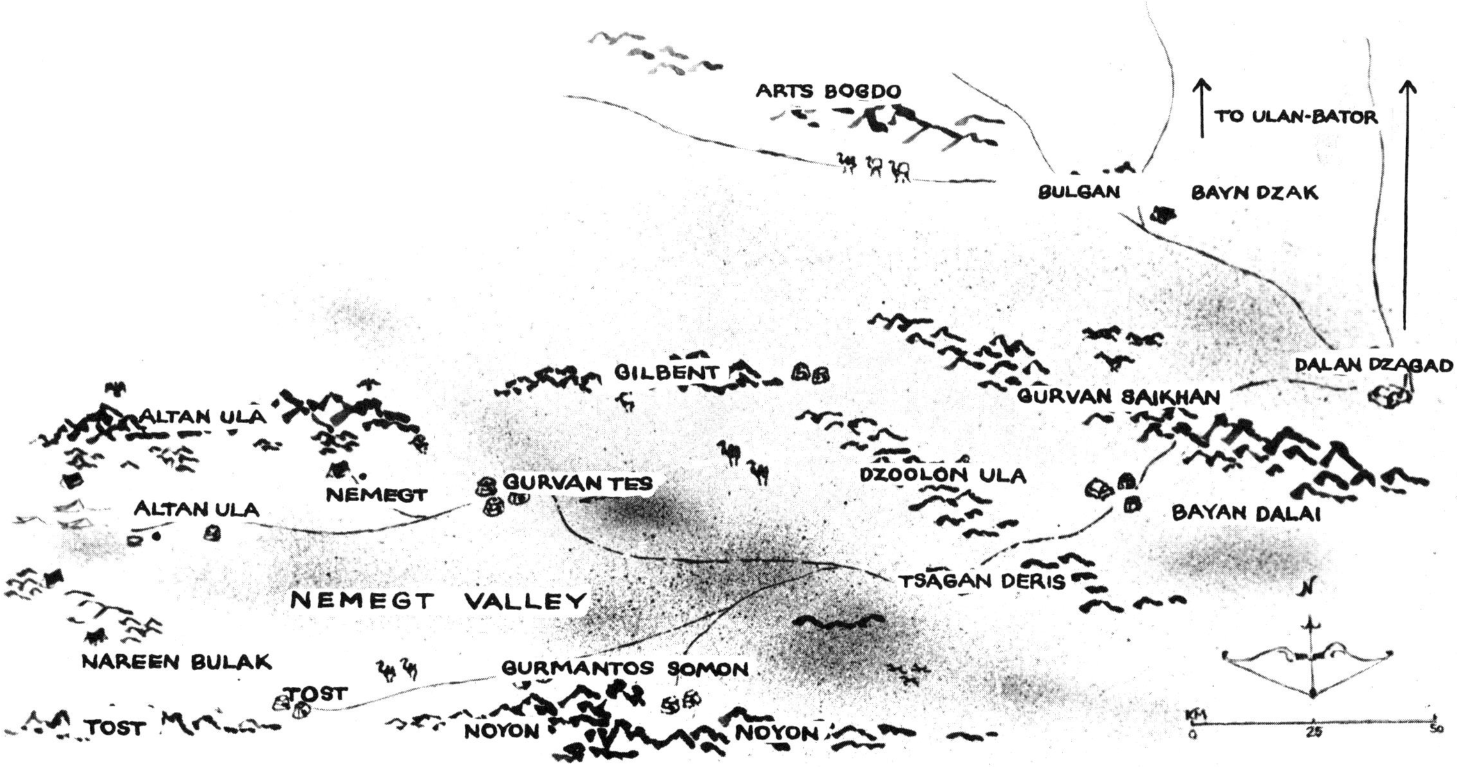

ARTS BOGDO
TO ULAN-BATOR
BULGAN
BAYN DZAK
GILBENT
DALAN DZAGAD
ALTAN ULA
GURVAN SAIKHAN
NEMEGT
GURVAN TES
DZOOLON ULA
ALTAN ULA
BAYAN DALAI
TSAGAN DERIS
NEMEGT VALLEY
N
NAREEN BULAK
GURMANTOS SOMON
TOST
TOST
NOYON
NOYON
KM
0
25
50

1. Introduction

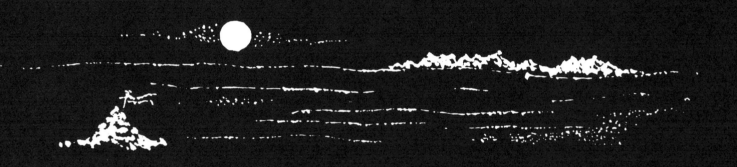

Introduction

I still distinctly remember the day I first heard about paleontological expeditions to the Gobi Desert. It was in 1946, and I was a student of paleontology under Professor Roman Kozłowski at the University of Warsaw. Warsaw was still in ruins after the war, and the lectures took place in a small room in Professor Kozłowski's own flat on Wilcza Street. There was a blackboard on the wall, painted by the Professor himself, a small table in front of it doing duty as a lectern, two tables with eight chairs around them, and, against the back wall, a bookshelf, the nucleus of a new library for the Institute of Paleontology at the University which had been totally destroyed in the war. The lectures, the first of their kind since the war, were attended by a group of six: two geology students, three zoologists, and myself, then a budding paleontologist. Despite the primitive conditions, it was a great adventure for all of us. Coming into the room, we would find pictures of the fossils to be discussed artistically drawn in colored chalk on the blackboard. The lectures themselves were fascinating and well delivered, though the Professor spoke so fast as to give writer's cramp to anyone trying to take notes.

We arrived one morning to find drawn on the blackboard the skulls of two mammals with the strange names *Deltatheridium* and *Zalambdalestes*. These were the earliest placental mammals known, about 95 million years old, discovered in 1925 in Cretaceous deposits by an American expedition to the Gobi Desert. It was then that I learned for the first time that the Gobi was a veritable Eldorado for paleontologists. For the last 135 million years of Earth history Central Asia, including what is today the Gobi Desert, has never been covered by the sea. In those early days, however, the Gobi was no desert; the Asian coastline was then entirely different and the Gobi region lay near the sea and enjoyed a moist climate. The shores of its large rivers and lakes were inhabited by dinosaurs, the gigantic reptiles of the Mesozoic era. On the steppes at some distance from the great waters, where the climate was drier, small, primitive mammals could already be found playing a very modest role in the continental fauna of the time, of which the dinosaur was the undisputed king.

I was fascinated by these American expeditions to the Gobi Desert, which had been described by the Professor in great detail. I began to search various libraries for more information and came up with an account of these famous expeditions by Anna Gadomska published in 1935 in the Polish periodical "Kosmos." I learned from this that the American Museum of Natural History in New York sent five successive paleontological expeditions to the Gobi between 1922 and 1930, headed by the well-known explorer Roy Chapman Andrews. I also read the Polish translation of Andrews' book, "To the Ends of the Earth," which included an account of his stay in the Gobi, and managed to unearth his reports in the National Geographic Magazine and a few other popular articles.

However, not in my wildest dreams did I expect that I, too, would some day go there, and that only 16 years later I would be organizing a Polish-Mongolian paleontological expedition to the Gobi Desert. Had some soothsayer foretold it at the time, I would certainly not have believed him. In any case, I spoke to no soothsayers—they were out of fashion just then—and the only paleontological expeditions in which we could participate were to Świętokrzyskie Góry [Holy-Cross Mountains] in Poland.

In the years 1948–49 news reached us that Soviet paleontologists were already going to the Gobi. These expeditions had been organized by the Paleontological Institute of the Soviet Academy of Sciences, and Soviet paleontologists had actually visited the Gobi on three separate occasions (in 1946, 1948, and 1949) to conduct large-scale digs in search of large dinosaur skeletons. We in Poland followed their work with great interest, hoping against hope one day to be able to participate in such an expedition ourselves.

In 1955 I was able to go to Moscow for the first time in my life and had the opportunity of visiting the Paleontological Institute of the Soviet Academy of Sciences, as well as the museum attached to that Institute. I saw mounted skeletons of large carnivorous dinosaurs of the Cretaceous—*Tarbosaurus*—and of the enormous herbivorous dinosaur Saurolophus, which had been found in the Nemegt Basin in the southern Gobi. Anatole K. Rozhdestvensky, who had taken part in the 1948 and 1949 Soviet expeditions, showed me large albums of photographs taken on the expeditions and gave me a detailed account of his stay and adventures in the Gobi. These details were already familiar to me from his book "In the Footsteps of Dinosaurs in the Gobi Desert," which I had read in Polish translation.

Sending Polish paleontological expeditions to Mongolia came under serious consideration in March 1961 at the Warsaw convention of representatives of Academies of Sciences of Eastern Bloc Countries. The participants included the Chairman of the recently established Academy of Sciences of the Mongolian People's Republic, Professor Shyrendyb. It was then suggested by the Paleozoological Institute of the Polish Academy of Sciences in Warsaw that joint Polish-Mongolian expeditions be sent to the Gobi Desert; the suggestion came from the dean of Polish paleontologists, Professor Roman Kozłowski, and met with a positive response from the authorities of both academies. A delegation from the Governing Board of the Polish Academy of Sciences (including Professor Kozłowski) then visited Ulan-Bator in September 1962 to discuss the details, and an agreement to organize joint, three-year Polish-Mongolian paleontological expeditions to the Gobi Desert was signed. On the delegation's return to Poland, I was instructed by the Polish Academy of Sciences to organize the expeditions and to take charge of their scientific side.

It was decided that the first expedition should leave for Mongolia as early as 1963 to make a preliminary survey. In order for field work to begin in June, all the equipment and supplies had to be sent from Poland by rail in February, to be collected by the expedition personnel arriving in Ulan-Bator in May. The date was October 1962, which meant that we had less than four months to prepare the scientific strategy for the expedition and to buy and crate the equipment and ship it to Ulan-Bator.

This attempt to organize paleontological expeditions to Mongolia took not just energy and enthusiasm but a great deal of courage and even foolhardiness as well. The preparation of a scientific expedition which is to conduct excavations dozens or even hundreds of miles away from the nearest human habitation demands a high degree of technical competence and extensive experience. The first difficulty was to draw up the scientific work plan. The research experience of the Paleozoological Institute of the Polish Academy of Sciences, which had been put in charge of the preparations, lay almost exclusively in the field of invertebrate paleontology. The only vertebrate fossils found and studied up to then in Poland were Devonian and Tertiary fish, along with mammals from the late Tertiary (Pliocene) and Quaternary, that is to say animals geologically very young. In the Cretaceous, when gigantic dinosaurs roamed what is now the Gobi Desert, the area that is now Poland lay beneath the sea. Marly and calcareous deposits dating from this period may be found in Poland in several places, for example in the vicinity of Kazimierz on the Vistula. They contain skeletons of such marine invertebrates as pelecypods, ammonites, and belemnites, but of course no skeletons of terrestrial dinosaurs. The area remained covered by the sea during the subsequent Tertiary period, whereas the Gobi was then dry land inhabited by various groups of mammals.

Thus thorough scientific study was needed to prepare for excavating skeletons of large, 95- to 75-million-year-old dinosaurs and of primitive mammals from the Cretaceous and the first half of the Tertiary. The fossils native to Poland, with which we were thoroughly familiar, were of no help in this respect. Not a single museum in Poland could boast of a single skeleton, a single bone, or even so much as a cast of a dinosaur skeleton from which to study the anatomy of these animals. As for mammals, no skulls older than the Pliocene were available. Thus, all the available information was only theoretical, and even there serious difficulties were encountered. Numerous fundamental works on dinosaurs and primitive mammals had perished in museums devastated by the war and had to be imported on microfilm. Preliminary inspection showed that the basic literature on the Gobi's geological structure and fossil animals comprised over 200 items. These were grouped according to subject and distributed among the prospective members of the expeditions. In winter, before the departure of each expedition, seminars lasting several days were held with endless discussions on the geological structure of the Gobi and the fossils so far described. A worldwide list was drawn up of all Cretaceous dinosaur species, with special emphasis on the species found in Asia. This was done in order to be in a position to compare the Gobi fauna with that of other continents and to get an idea of the forms which might be newly discovered in the Gobi. Similar lists were also prepared for certain groups of mammals.

Technical preparations for the expedition were carried out at the same time. I made a preliminary list of equipment needed and was forced to the conclusion that the technical arrangements would have to be left to an expert if the expedition were to succeed. We had less than four months in which to buy several thousand items, none of which we could afford to find missing while out in the field.

How was such an expert to be found? It appeared at first that there was no such profession, at least not one taught in any university or technical school. I was nevertheless convinced that a suitable man could be found if we looked hard enough. And indeed he was; he turned out to be Maciej Kuczyński, an engineering graduate. He had been the technician in charge of the Spitzbergen Expedition organized several years before within the framework of the International Geophysical Year, had led speleological expeditions to Cuba and Bulgaria, and had taken part in a number of other expeditions.

He began by drawing up a complete list of equipment to be purchased. The list, which took several days to compile and ran to over 20 pages, included, on the one hand, such items as a truck, 55-gallon gasoline drums and a 3 months' supply of gasoline and on the other, such items as pens and drawing ink, tracing and millimeter paper, drawing pins and nails, hooks and screws, carpenter's and ironmonger's tools, tents, pots, spoons, knives and forks, mess tins and complete camping equipment, table awnings, pegs for erecting field kitchens, sets of spatulas and brushes, polystyrene cement for bones and acetone for dissolving the polystyrene, sacked plaster, sacked sawdust for packing bones, bolts of gauze in which to wrap up the bones and prepare plaster dummies, a medicine chest to meet all possible contingencies, a varied three months' supply of food, flashlights and candles, an automobile repair kit, a mechanical trailer with cable ropes, crates and boards, padded clothing for cold weather and light clothing for hot, stout walking boots, knapsacks, sleeping bags, inflatable mattresses, collapsible beds, a typewriter and hundreds of other indispensable items.

A large room, ordinarily used to receive collections arriving at the Institute, soon turned into a veritable general store. At the beginning of February a "Star"-25 truck bought by the Institute at the Starachowice factory made its first entry into the Institute courtyard, and by the end of the month all the equipment had been crated and sent to Ulan-Bator by rail. In May 1963 a group of five Polish paleontologists, members of the first joint Polish-Mongolian survey expedition, left Warsaw by plane for Ulan-Bator. The first of our Gobi expeditions had begun.

2. Dinosaurs and Mammals

11 Dinosaurs and Mammals

Dinosaurs, our "game" in the Gobi Desert, are among the most popular animals of past geological time. Their history begins in the Triassic, i. e., some 220 million years ago. The dinosaurs were undisputed masters of the Earth's continents throughout the 155 million years of the Mesozoic era, i. e., during the Triassic, Jurassic, and Cretaceous periods. They became extinct toward the end of the Cretaceous, 70 million years ago.

It is widely believed that dinosaurs were the biggest animals of all time. It is true that they did include gigantic forms—in fact, the biggest animals ever to walk the earth, such as the famous *Brachiosaurus,* as tall as a modern three-story building, and the mighty *Brontosaurus,* over 60 feet long. However, they hold the size record only on land. In the sea today there are certain whale species bigger and heavier than the largest of dinosaurs. Moreover, not all dinosaur species were big. Alongside the gigantic forms there were many medium-sized species 3 to 6 feet long, and some were even quite small, about the size of a turkey.

The dinosaurs were reptiles. In our present-day fauna the reptiles are a relatively small group embracing the lizards, snakes, turtles, crocodiles, and the so-called rhynchocephalians, of which only one vanishing species, the New Zealand tuatara, still survives. In past geological time, however, reptiles were extremely abundant and appeared in an astonishingly large number of different forms.

The oldest reptiles date from the Carboniferous period in the second half of the Paleozoic era, i.e., from about 330 million years ago. These earliest reptiles bear the name cotylosaurs. They were heavy, awkward animals, with short, thick limbs spread out to their sides. In the late Carboniferous (Pennsylvanian) the cotylosaurs gave rise to the first mammal-like reptiles; the latter were to play an important role in the further development of the vertebrates, since they were the immediate forerunners of the first mammals—which appeared in the Triassic, early in the Mesozoic. The mammals will be discussed a little later in this chapter; for the time being, we shall try to follow the subsequent development of the reptiles in the Mesozoic.

The Mesozoic era is known as the Age of Reptiles. The variety of forms in the various reptile groups which began to reign in the Triassic all over our planet, in the oceans and fresh water lakes, on the continents and in the air above, is simply astounding and is comparable only to the abundance of mammal forms in our present-day fauna. The Mesozoic reptiles which adapted themselves to living in the sea included the sharklike ichthyosaurs, the large plesiosaurs (which resembled a turtle with a snake threaded through it), and, finally, those mighty, predatory sea lizards, the mosasaurs. But the real masters of the Mesozoic are considered the archosaurs or Ruling Reptiles, the group which includes the dinosaurs.

The origins of this group can be traced to the Triassic, when the modest thecodonts, descendants of the cotylosaurs, made their appearance on dry land. These were relatively small, predatory animals looking somewhat like lizards, with very long, narrow skulls and sharp teeth. The thecodonts gave rise, on the one hand, to the birds, and on the other, to the crocodiles, the flying reptiles, and the dinosaurs. These three commonly descended reptile stocks taken together form the Ruling Reptile group. There is no need to describe the crocodiles, since they are the only line of Ruling Reptiles still extant and can be seen in any zoo. The flying reptiles, or pterosaurs, are an interesting group of extinct forms which took to the air, but whose flight was based on a principle different from that of birds. While birds use feathers and bats a flying membrane spread between the elongated finger-bones of the hand as their flight plane, pterosaurs were able to fly by using a skin membrane spread out between the elongated phalanges of the fourth finger and the side of the body.

The Ruling Reptiles of greatest interest to us are the dinosaurs. The name "dinosaur"—meaning "terrible lizard"—was coined in the middle of the last century by Richard Owen, the first to study dinosaurs in Great Britain. At that time the term was used to denote all the large reptiles of the Mesozoic era. However, more thorough study of dinosaur anatomy in the United States and Europe during the second half of the 19th century showed that the name dinosaur, as then used, covered two groups not directly related: the Saurischia [those with triradiate (reptilelike) pelvises] and the Ornithischia [those with tetraradiate (birdlike) type of pelvises]. Both these groups were descended from thecodonts, but they differed in subsequent development and are no more related to one another than either is to the crocodiles or pterosaurs. The main difference between the two dinosaur orders is in the structure of the pelvic girdle, which in all reptiles consists of three parts: the ilium, the ischium, and the pubic bone. In the Saurischia all three bones are at an angle to each other, making the pelvis triradiate. In the Ornithischia the pubic bone carries a backward stump parallel to the ischium, so that the pelvis becomes tetraradiate and resembles a bird's pelvis in general form. This is an important anatomical difference, by which each dinosaur may be classified as belonging to one or the other of the two orders.

Both orders inherited the thecodont tendency to bipedalism, manifested by the fact that the hind limbs are, with very few exceptions, longer than the forelimbs. Both orders include bipedal dinosaurs, and the quadrupedal ones got to be that way as a result of secondary development. Bipedal dinosaurs developed long, very muscular tails, which acted as a prop and helped the animal keep its balance. The entire weight of the body rested on the pelvis and the hind limbs, which were vertical and able to shift under the pelvis.

The Saurischia included two groups: a carnivorous one, the theropods; and a herbivorous one, the sauropods. The theropods branched off from the carnivorous thecodonts as far back as the Triassic, and this group preserved many more thecodont features than did other dinosaurs. All theropods were bipedal, walking semierect on their hind limbs. The group included the dreadful predatory carnosaurs. In certain species of Cretaceous carnosaurs the forelimbs underwent a considerable reduction in size. The theropods also include the so-called ostrich-like dinosaurs (or ornithomimids) that outwardly resembled birds, in particular ostriches. These were bipedal animals whose long necks bore a small head equipped with a toothless beak, and which had long, prehensile forelimbs.

The second group of Saurischia—the sauropods—included herbivorous, quadrupedal forms. Their ancestors were the Triassic prosauropods, which still walked semierect. Among the sauropods were the biggest land animals of all times, the famous Jurassic giants *Brachiosaurus, Brontosaurus,* and *Diplodocus.* They were heavy quadrupeds with a massive trunk supported on pillarlike limbs; their long necks bore a small head, and their tail was very long and whiplike. The sauropods' greatest development took place in the Jurassic; complete skeletons have been found in Jurassic sedimentary rocks in the United States, East Africa, and elsewhere. During the Cretaceous, they became less common but some representatives of the group survived till the end of that period.

The other dinosaur order, the Ornithischia, was much more varied in its forms, both in appearance and in adaptation to different environments. Some were bipeds, while others were quadrupeds, the former using their forelimbs as aids in walking. All known representatives of the order were herbivores. They included large animals, some over 30 feet long, but none were giants comparable to the sauropods. The ornithischians were rare at the beginning of dinosaurian history, but recently some well-preserved late Triassic ornithischian remains have been found in South Africa. A considerable radiation of the group was certainly taking place during the Jurassic; their remains from this period are, however, rare. The best known Jurassic ornithischians are the stegosaurs, peculiar-looking quadrupedal dinosaurs up to 30 feet long. Stegosaurs had two rows of vertical bony plates of enormous size along their backs and also bore two pairs of very large spikes on their very muscular tails. The head and brain sizes of the animals were astoundingly small, the brain being about as large as a walnut and weighing no more than four ounces, while the weight of the animal is estimated at 10 tons.

The best known of the bipedal Ornithischia are the iguanodons, which are common in Lower Cretaceous formations in Europe.

However, the greatest development of the ornithischian dinosaurs took place during the second half of the Cretaceous. The ornithischian biped group—the ornithopods—displayed at that time an abundance of different forms. These include the so-called duckbilled dinosaurs, in which the muzzle was covered with a horny sheath much like a duck's bill but which had sharp teeth on the sides of the jaws. These duckbilled dinosaurs included the saurolophs, complete skeletons of which have been found by Soviet paleontologists in the Gobi Desert, and also the pachycephalosaurs, which were found in Upper Cretaceous formations in North America. As a means of passive defense this group possessed a very thick-boned skull-roof forming a dome nearly a foot thick in some species. This thickening of the cranial bones produced a highly arched forehead over the eyes, giving the pachycephalosaurs an "intellectual" appearance. This appearance was a mere illusion, however, since the brain hidden under the thick bone layer was very small, just like that of any other Mesozoic reptile. The pachycephalosaurs, like all other dinosaurs, were animals with a very low level of intelligence.

The two other dinosaur groups of the order Ornithischia—the groups typical of the second half of the Cretaceous—were the armored dinosaurs and the horned dinosaurs. Both were quadrupedal and thus were slow runners unable to flee from predatory dinosaurs. Accordingly, many passive means of defense were developed. The armored reptile group called ankylosaurs—sometimes also "reptilian tanks"—is particularly noteworthy in this respect. These were large animals reaching lengths of 15 feet. Their body was flat, squat, and wide, and was covered on the outside by a mosaic of very thick bony plates of various sizes. The wide, flat skull, like the rest of the body, was also covered with a supplementary layer of bony plates. Certain species developed rows of large spikes along the sides of the trunk. Thus on all sides of the body there was a protective armor so thick in some species that even the tyrannosaurs—the most dangerous beasts of prey of the time—could bite through it only with great difficulty.

The members of the horned-dinosaur group had other means of defense against their natural enemies. Skeletons of a small dinosaur, *Protoceratops*—thought to be the ancestor of this group—have been found preserved in sediments dating from the beginning of the second half of the Cretaceous on a site in the Gobi Desert called Bayn Dzak.* *Protoceratops* was a quadruped, and up to 5 feet long. Its skull was disproportionately large, with a huge, collarlike extension in the back covering the neck. The jaws were bill-shaped, much

---

*The locality was named Shabarakh Usu by U.S. paleontologists, this being the name of a small swampy lake in the vicinity.

15 Dinosaurs and Mammals

like the beak of a parrot. This ancestor of the horned dinosaurs did not as yet bear on its skull the horns so characteristic of the late Cretaceous representatives of the group. One of the best known of those of the horned dinosaurs which lived toward the end of the Cretaceous is the mighty *Triceratops*. This was a much larger animal than its *Protoceratops* ancestor, being 20 feet long and 8 feet high. It had three horns on the front of the skull: two bony horns over the eyes and one horn on the nose. The nape of the neck was shielded by a large, collarlike bony crest which was an extension of the skull.

The dinosaurs, after holding undisputed sway on land throughout the Mesozoic era, became extinct toward the end of the Cretaceous. The following period in Earth history—the Tertiary, which opened the Cenozoic era—is characterized by a radical change in the forms of life on land. The large dinosaurs disappeared, to be succeeded by rapidly developing groups of mammals, in particular by the placental mammals.

If we compare the developmental history of the mammals with that of the dinosaurs, we can see how different the evolutionary paths of different groups of animals can be. It is hard to believe that the dinosaurs and mammals could have appeared on Earth at almost the same time, back in the Triassic. However, while the dinosaurs immediately began to differentiate very rapidly and took over all the continents of our planet during the Mesozoic era, the mammals proceeded to develop very slowly, branching out into but a few orders, and those included only small forms—mouse-sized, rat-sized, or at most cat-sized—whose contribution to the Mesozoic fauna was insignificant. Thus the mammals, which appeared toward the end of the Triassic, i.e., about 185 million years ago, developed very slowly for the first 115 million years of their existence. It was only 70 million years ago, when the Cretaceous merged into the Tertiary, that a radical change occurred in the forms of life on Earth, the most radical change in biohistory so far. The dinosaurs disappeared completely from the surface of the earth and were succeeded by the mammals, which from the start of the Tertiary period began an explosive differentiation. Just as the reptiles had done in the Mesozoic, so the mammals in the Tertiary took over everywhere on land—but not on land alone. Some mammals, such as whales, dolphins, sirenians, and seals, adapted themselves to the sea, just as ichthyosaurs, plesiosaurs, and mosasaurs had done 100 million years before. Others, such as bats, learned to fly, using a principle similar to that of the flying reptiles of yore, i.e., by forming a flight membrane. The variety of mammal shapes and adaptations in the Tertiary may be compared with the various adaptations achieved by the dinosaurs; there are some striking analogies, such as the external similarity between the armored reptiles and the South American armadillos.

Even though mammals are descended from reptiles, the differences are very considerable. The reptile body is covered with scales, the mammal body with hair. The blood temperature of reptiles is variable, that of mammals is constant. Reptiles are oviparous, while mammals are viviparous (except for the egg-laying monotremes in Australia); mammals suckle and care for their young after birth, reptiles do not. The skeletal structures of reptiles and mammals also show wide differences. The ossification of the skeleton is more advanced in mammals than in reptiles. In reptiles the bones keep growing throughout life, whereas the cartilaginous endings of mammalian bones ossify at a certain age and bone growth stops. The mammal skull is joined to the backbone by two occipital condyles, while reptiles have only one such condyle (it should be pointed out, however, that the mammals' immediate ancestors, the mammal-like reptiles, already had two occipital condyles). One typical feature of the mammal skeleton is the smaller number of bones in the skull, since many have joined together. The mammalian braincase is relatively large, because of the large size of the brain. Reptile teeth are peglike and similar all along the jaw, whereas mammal teeth are differentiated into incisors, canines, premolars, and molars. Finally, reptiles have a differently shaped jaw joint and a different jaw structure. The reptile lower jaw consists of six and sometimes nine bones on each side (the mammal-like reptiles, the forerunners of the mammals, had fewer bones in the jaw—only three in each half of the lower jaw). The lower jaw of mammals, on the other hand, consists only of one bone, called the dentary, and the jaw joint has a different structure. Detailed paleontological and comparative anatomical studies have shown that the bones forming the jaw joint in reptiles, i.e., the quadrate bone (which is a part of the skull) and the articular bone (on the side of the jaw), were very small in mammal-like reptiles and had shifted to the vicinity of the ear; in mammals they form part of the ear and become auditory ossicles. Thus, where the reptile ear contains only one bone to transmit auditory vibrations, mammals have three. These are the stapes or stirrup bone, which corresponds to the single ear bone of reptiles and amphibians, and two additional ones: the incus or anvil, which originates from the quadrate bone, and the malleus or hammer, which derives from the articular bone. Since the bones constituting the jaw joint in reptiles form part of the ear in mammals, the latter have developed a new jaw joint—between the dentary bone forming the jaw and the squamosal bone in the skull. Recent discoveries of primitive Triassic mammals have shown that, as might have been expected, some of the oldest mammals had two joints on each side of the jaw: a reptilian and a typically mammalian one. The mammal-like reptiles of the Triassic are so

17 Dinosaurs and Mammals

close to the oldest mammals that it is very difficult to make a sharp distinction between the two on the basis of the available fossil material alone and to identify each form conclusively as being still reptilian or already mammalian.

Triassic and Jurassic mammals are in general known only very sketchily, from single teeth or from jaw fragments that include teeth. Most belong to orders that became extinct during the Mesozoic. The best differentiated group of Mesozoic mammals comprises the multituberculates, whose oldest representatives appeared toward the end of the Jurassic and whose youngest date from the first half of the Tertiary (Eocene). These were herbivorous mammals with skulls at first glance like those of today's rodents. Just as in rodents, the canines had disappeared and a diastema had appeared at their place in the jaw between the elongated, chisel-shaped incisors (suited to chopping up hard plants) and the cheek teeth. The molars were elongated and covered with numerous cusps. Multituberculates were highly specialized in the Cretaceous and early Tertiary, and reached large sizes for mammals of their day. The skulls of the largest known multituberculates are up to 7 inches long. The group died out without descendants in the first half of the Tertiary, probably because of an inability to compete with the simultaneously developing placental mammals.

Unfortunately, monotremes—the only living egg-laying mammals are almost unknown as fossils. But without doubt this is a very primitive group, which must date back to the Triassic. Fossil marsupials from the Mesozoic, however, are known.

The placental (or eutherian) mammals are the most numerous and vigorously developing group of contemporary mammals. Their various orders have developed during the last 70 million years of Earth history, i.e., since the beginning of the Cenozoic era. However, already at the beginning of this era in the early Tertiary, different groups of placentals were strongly differentiated, indicating that the group must have had a long history in the Cretaceous and possibly even earlier. The first Mesozoic placentals were discovered by American paleontologists in Upper Cretaceous deposits in the Gobi Desert. During the more than forty years that have elapsed since, in particular during the last decade, an intensive search has gone on, mainly in the United States, for placental mammals in Mesozoic formations. By now, fragmentary placental fossils have been found in North America even in Lower Cretaceous deposits, and one incomplete lower jaw has been found in the Lower Cretaceous of Manchuria.

Most of the Cretaceous mammal fossils found as a result of this intense search are very incomplete. They may consist of isolated teeth, and of fragmentary lower jaws, or more rarely of upper jaws. The best-preserved Cretaceous mammal material found so far—

complete or almost complete, well-preserved skulls—has been from the Gobi Desert. It was therefore only natural that the Polish-Mongolian paleontological expeditions to the Gobi should make a special effort to find such mammals, so that a unique collection of skulls of primitive placental mammals and multituberculates, of great scientific interest, could be assembled.

On our expeditions to the Gobi we investigated outcrops of the lower part of the Upper Cretaceous—deposits containing small mammals accompanied by land dinosaurs (*Protoceratops* and armored dinosaurs). We also dug in outcrops of the uppermost Cretaceous alluvial deposits in which skeletons of large dinosaurs are preserved, as well as in Tertiary (Paleocene) outcrops, which no longer contain any dinosaurs but have only mammal fossils. We were thus able to come into direct contact with one of the greatest enigmas of paleontology, namely the changes in the fauna at the end of the Cretaceous and beginning of the Tertiary and the extinction of the dinosaurs. A few words on this subject may not be out of place here. No paleontologist can fail to be struck by the strange extinction of the enormous dynasty of the dinosaurs, which held undisputed sway over the earth throughout the 155 million years of the Mesozoic era. The reasons behind this extinction form one of the most difficult problems of paleontology, one for which no satisfactory answer has yet been found.

It is generally thought that the dinosaur's most vigorous development took place in the Jurassic, since the skeletons of the largest ones known—the giant sauropods—date from this period. If, however, we examine the number of individual dinosaur forms and their adaptations in the different periods, we see that a much greater differentiation took place in the Late Cretaceous. Of the Saurischia, the sauropods and the big carnivores (the theropods) survived to the end of the Cretaceous, while in addition the ostrich-like dinosaurs developed in the Late Cretaceous. Of the Ornithischia, the stegosaurs failed to survive to the Cretaceous, but new groups developed in the Late Cretaceous: the armored, the horned, and the duckbilled dinosaurs. Thus the extinction of the dinosaurs toward the end of the Cretaceous, far from coinciding with their decline, coincided with their greatest development.

In the second half of the 19th century it was fashionable to believe in the theory of the racial senescence. The development of a race (or generally a group of animals) was compared with that of an individual, and just as any individual must pass through stages of youth, maturity, senescence, and death, so the race, too, was supposed to have a certain stock of vital energy that became depleted as the race developed. The exhaustion of this vital energy would then correspond to the extinction of the group. Quite apart from the fact that the concept of "vital energy" cannot in any way be rationally justified, these theories could not be maintained, simply because they contradicted the facts furnished by paleontology. In fact, comparison of the developmental histories of the dinosaurs and the mammals, both of which appeared on earth almost at the same time but evolved altogether differently, indicates that the evolution of animal groups knows no stages of youth, maturity, and senescence to compare with the similar stages in the lives of individuals.

According to the Georgian paleontologist Davitashvili, the dinosaurs' extinction was the result of the competition ("struggle for existence") against the contemporary mammals. The Mesozoic mammals, though much smaller than the dinosaurs, were very much their superiors in brain development and intelligence. Thus they could, for example, have contributed to the extinction of the dinosaurs through eating their eggs, which were unguarded on the ground. This view, however, is not fully convincing. The Mesozoic mammals lived in plateaus, steppes, or forests, far from major bodies of water. Now, the armored and the horned dinosaurs, which were adapted to living on land, may possibly have shared these territories with the mammals, but most dinosaurs—the sauropods and most of the ornithopods (e.g., all the duckbilled dinosaurs)—lived probably far from the haunts of mammals. These large reptiles led presumably amphibious lives and dwelt in large, shallow lakes inaccessible to mammals. Therefore mammals could not have brought about the extinction of all the dinosaurs, since their respective environments as a rule did not overlap.

According to other theories, the dinosaurs died out as a result of changes in climate and in other ecological factors; it is known, however, that the advent of a cold climate, however abrupt, never takes place all over the globe at the same time, nor from one year to the next, but rather extends over many thousands of years, during which animals are able to migrate to warmer lands.

Observation of the extinction of modern animal species, many of which have become extinct in historical times and others which will soon become so, indicate that many fall victim (to man) not only directly through thoughtless slaughter but also indirectly through destruction of their natural environments. The felling of the virgin forests of Europe in historical times, for example, has caused the extinction of many mammal species, including the aurochs (*Bos primigenius*)—the ancestor of the cow—and the bison (which has survived only on special reservations). Development of pasture lands in various areas produces a decrease in water-bird populations, and is sure to lead to the extinction of many individual species. It can already be predicted which species of African mammals will become extinct (or survive only on reservations) as a result of the advance of civilization on that continent.

Could it be that the disappearance of certain environments, here not through human activity but because of physical factors, was responsible for the extinction of the dinosaurs? This was suggested to me in the course of a conversation by the geologist, Professor Stanisław Dzułynski, who pointed out that types of continental deposits typical of the Mesozoic era no longer appear in the Tertiary. During the Mesozoic, the earth's land surface was relatively flat. There were enormous coastal lowlands and numerous inland lakes, occasionally very extensive but very shallow, their maximum depths being a little over 30 feet. Such basins were an ideal environment for large dinosaurs. Big animals need a lot of space in which to live. It is well known that the island representatives of continental species are smaller than the mainland ones. Large animals cannot subsist on mountainous terrain cut through with narrow riverbeds. It is unlikely that the large species of dinosaurs would be able to find a suitable environment on the surface of the earth as it is today. Such environments probably disappeared at the beginning of the Tertiary, due to the Alpine mountain-making revolution. The disappearance of shallow lakes and wide open coastal spaces would have caused the extinction of the herbivorous dinosaurs, and thus of the carnivorous dinosaurs as well, by depriving them of their prey.

This explanation for the extinction of the dinosaurs is offered only as a hypothesis, one which may well be valid but which still needs further verification. Such verification would require a detailed sedimentological investigation of the continental Mesozoic and Tertiary sediments in every area in which mammal and dinosaur bones are known to have been preserved.

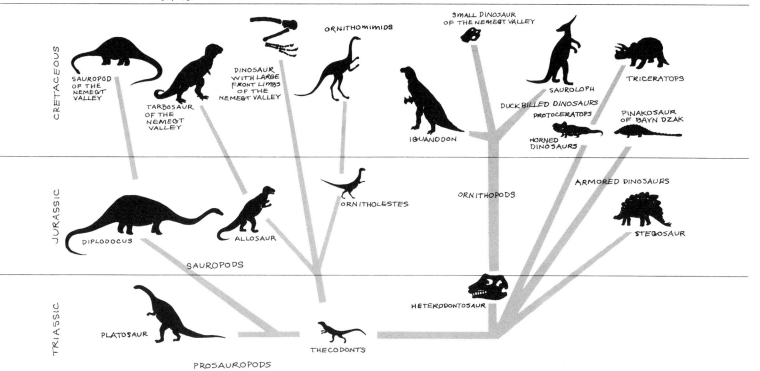

3. A Bit of History

A Bit of History

By the end of the 19th century paleontological investigation was fairly advanced in the United States and in various European countries. A subject widely discussed at that time was the problem of the birthplace of the placental mammals. A rich placental fauna had been found in Europe and the United States in deposits from the Paleocene, at the base of the Tertiary, but not in the underlying Cretaceous formations. Now, the extent of differentiation of the placentals in the Paleocene deposits indicated that these animals must have existed for a long time prior to the Tertiary. It was therefore believed that the mammals had developed on some one particular continent during the Cretaceous and had then migrated to Europe and North America at the beginning of the Tertiary.

In 1900 the eminent American paleontologist Henry Fairfield Osborn suggested that the placental mammals' pre-Tertiary development had taken place in Central Asia, at that time virgin soil for paleontology.

Twenty years later, the problem of the origin of the placental mammals was taken up again at the Museum of Natural History in New York. This renewed interest was probably due in part to new paleontological discoveries in Asia—a Russian paleontologist, Borissyak, had located and described an interesting Tertiary mammal fauna in Kazakhstan. The geological structure of Mongolia indicated that deposits of this age would probably be encountered in the Gobi Desert. It was also thought that the Gobi would reveal some interesting Cretaceous fossils.

After two years of preparation, the first American expedition, led by the well-known explorer Roy Chapman Andrews, left New York in 1922. From 1922 to 1930, five successive American expeditions explored what is now the southern part of the Mongolian People's Republic, along with Inner Mongolian territory which now forms part of China. These expeditions were organized on a grand scale, the largest comprising forty participants. The main purpose was paleontological, but topographers, geographers, geologists, archeologists, botanists, and zoologists also took part. The main base of the expedition was in Peking. The facilities available in the Gobi Desert at that time were much more primitive than they are now: there was no railway through the desert (now there is the Irkutsk–Ulan-Bator–Peking line), no telephone service, and no airline. The American group traveled in a dozen or so small field cars, and was accompanied by camel caravans which set out ahead of it with the heavy equipment, including a supply of gasoline for the cars.

During these five years the Americans made rich collections of skulls and other dinosaur bones dating from the Cretaceous, along with numerous skulls and complete skeletons of Tertiary mammals. Transport difficulties ruled out large-scale excavations; thus no large dinosaur skeletons were brought back, but the paleontological material collected in this hitherto virgin territory was of enormous scientific value. Altogether more than 100 reptile and mammal skulls were found, as well as numerous fragments from skeletons of animals representing not merely new species, but even new genera and families. One of the highlights of the expedition was the discovery of the first dinosaur eggs ever seen, at Bayn Dzak in the Djadokhta beds of Cretaceous age. Paleontologists had long assumed that the dinosaurs, like the present-day reptiles, were oviparous, but even though many dinosaur skeletons had been discovered, no eggs had ever been found. The dinosaur eggs in the Bayn Dzak sandstone were relatively small, about 6 inches long, and had very well preserved, ornamented shells. They probably came from *Protoceratops,* numerous skeletons of which were also found in the same layers at Bayn Dzak. Today dinosaur eggs are more common; subsequent expeditions to the Gobi have brought back numerous nests of eggs from Bayn Dzak, and similar nests have been found elsewhere in the world, including China and southern France (Provence), where the Upper Cretaceous marls are full of broken eggshells.

The most valuable items brought back by the American expeditions are the skulls of primitive Cretaceous mammals found at Bayn Dzak in the same layers as the dinosaur eggs. The Americans put together ten incomplete skulls of small mammals, two belonging to multituberculate and the other eight to insectivores. This discovery was of great importance for science, since these were the first skulls of placental mammals ever found in Cretaceous deposits. These findings appeared to bear out Osborn's hypothesis that Central Asia was the center of development of placental mammals before the Tertiary. However, modern discoveries have since disproved the hypothesis, as many eutherian mammals have more recently been discovered in North American Cretaceous deposits, and a tooth was recently found in Cretaceous formations near Montpellier in France.

The next series of paleontological expeditions to the Gobi was organized by the Paleontological Institute of the USSR Academy of Sciences in Moscow. In 1941 the Scientific Commission of the Mongolian People's Republic asked the USSR Academy of Sciences to dispatch a paleontological expedition to Mongolia, but the outbreak of the war cut short the preparations. It was not until 1945 that the idea could be taken up again, and in 1946 the Paleontological Institute sent its first exploratory expedition to Mongolia, led by Professor Efremov, a well-known paleontologist and writer, other participants including the late Director of the Institute, Professor Orlov, and six other scientists and technicians. Even though it remained in the field for only two months, the expedition managed to make two traverses during that time, one of the southeastern Gobi and one of the southern Gobi, where a large cemetery of big dinosaurs was discovered in the Nemegt Valley Cretaceous deposits. The American expeditions had never reached this spot, so paleontologically speaking it was still virgin soil.

On the strength of these results two subsequent expeditions were organized, one in 1948 and one in 1949, with the aim of making large-scale excavations. Both were headed by Professor Efremov, as before. The 1948 Soviet group included 15 members and 14 laborers. Excavations were conducted south of Sayn Shand in the southeastern Gobi, in the Nemegt Valley and Bayn Dzak, in the southern Gobi, and finally more to the west, around lake Orok Noor. The most extensive excavations were in the Nemegt Valley, where valuable dinosaur material was found.

The 1949 expedition was on just as large a scale and involved 33 people (including laborers). Work was conducted first in western Mongolia—in the Valley of the Big Lakes; there, in the vicinity of Kobdo, the group discovered a rich deposit of Pliocene mammals, which it named Altan Teli, after the name of the old somon [an administrative unit corresponding to a county]. The expedition then continued work on areas investigated the previous year, i.e., mainly in the Nemegt Valley and in the southeastern Gobi.

In their three years of field work the Soviet paleontologists collected a rich store of skeletons, mainly of dinosaurs, filling 460 crates with their "monoliths" (bones together with rock coated over with plaster), the over-all weight coming to 120 tons. In the Nemegt Valley, the Upper Cretaceous outcrops at Nemegt, Tsagan Khushu, and Altan Ula yielded ten complete skeletons of carnivorous dinosaurs (*Tarbosaurus*) and duckbilled dinosaurs (*Saurolophus*). Just as interesting were the armored dinosaurs found at Bayn Shireh in the eastern Gobi and at Bayn Dzak in the southern Gobi. In addition to the dinosaurs, which have since been suitably prepared and mounted in the Moscow Paleontological Museum, the Soviet expeditions collected valuable mammal material from the Lower (Paleocene), Middle (Oligocene), and Upper Tertiary (Pliocene).

Nevertheless, despite the existence of these large collections assembled by Soviet and American expeditions at great cost in labor and money, we were quite certain that paleontological exploration of the Gobi was far from complete. Both the Cretaceous and the Tertiary deposits in this area are very rich in animal fossils, and only part of these have been discovered and described. We were firmly convinced that valuable new scientific material would be discovered by subsequent expeditions. Thus, as soon as the occasion presented itself, we embarked enthusiastically on organizing a third series of paleontological expeditions to the Gobi Desert—the Polish-Mongolian expeditions.

4. The First Reconnaissance Expedition and Subsequent Preparations

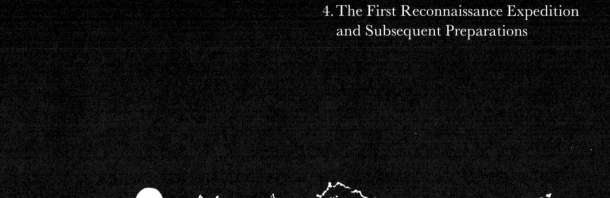

A Mongolian girl in the southern Gobi
Paleocene outcrops at Naran Bulak in the Nemegt Valley

31 Reconnaissance and Preparations

The five Polish participants on the first exploratory Polish-Mongolian paleontological expedition of the Gobi Desert left Poland in May, 1963, and returned to Warsaw in September. All came back in excellent spirits, sunburnt, somehow changed. The head of the expedition, Dr. Julian Kulczycki handed in his report. There were five Mongolian participants, including the paleontologists Dovchin and Dashzeveg who were due to arrive in Warsaw shortly for a three months' training course at the Institute.

At this juncture we should point out that there are no first names in Mongolian. The newborn child receives a surname which does double duty as first name. Thus, father and son have different surnames. A married woman does not take the name of her husband but keeps her own. The result is that each member of a family has a different name. Mongolians studying or working in European countries usually take first names, the common practice being to use the father's surname as one's own first name. First names are not used in Mongolia itself.

It had not been the purpose of this first reconnaissance expedition to conduct any excavations. Its task was to survey the southeastern and southern Gobi, to familiarize itself with the Cretaceous and Tertiary outcrops, to locate the major accumulations of fossil bones, and to establish working locations for the succeeding expeditions. The members had gone out in two trucks, a Polish Star 25 and a Mongolian Gaz 63, and had covered more than 5000 miles. The results of the trip were of great help in planning the following expedition. We now knew where the biggest accumulations of bones were and how to reach them; moreover, we were now familiar with conditions on the Gobi, and so could select the excavatory and camping equipment and the required foodstuffs in the best way possible. We had found, for example, that our [two-wheel-drive] Star 25 was unfit for use under desert conditions. It gave excellent service on the vast steppe and semidesert stretches, but in the Nemegt Valley, for example, where it often had to be driven in the sayrs—the dry seasonal streambeds—or over sand-dunes, it would often bog down and have to be dug out, giving endless trouble. It was therefore decided to use another type of vehicle for the 1964 expedition, and the Star 25, together with all the expedition's gear, was left behind in Ulan-Bator for the winter. We were to use this vehicle the following year on the less exacting routes and in hauling the collections to Ulan-Bator, but for desert driving we needed something more suitable. Here the "Star" truck factory in Starachowice came to our aid, putting their excellent six-wheel-drive model "Star 66" at our disposal for the 1964 expedition. This vehicle can be used anywhere, even for driving over sand dunes, no matter how difficult the terrain.

We had also gained some experience with camping tents. The Gobi

Desert is famous for its strong winds, seasonal sandstorms, and hurricanes. It was found that the Polish-made tents taken on the reconnaissance expedition could not withstand these winds; the 1964 expedition would have to use a different kind. Mr. Kuczyński designed a new type of tent, based on those used in Mongolia. This had two stout metal uprights, joined at the top by a thick metal crossbar. There was no flooring, in view of the high temperatures out on the Gobi. At the hottest time of day the tent walls could be rolled up on all sides, to ensure circulation of air and to reduce the temperature of the sun-scorched roof. This procedure had its drawbacks as well: any small object left lying about was liable to be blown out and could then conveniently be watched sailing through the air high above the tops of the tents. On the 1964 and 1965 expeditions we found the performance of this type of tent in the Mongolian climate satisfactory, as a rule. During real sandstorms, however, such as we experienced two or three times on each expedition, neither Mongolian tents nor our own could withstand the force of the wind. The only tents fit to stand up to a Mongolian hurricane are the round portable *yurts* used by Mongolian nomads, whose design is the result of thousands of years' experience; we could not, however, take these along, since they are too heavy and too awkward to set up and take down.

Even though it did no digging, the reconnaissance expedition collected a few cases of fossils from surface outcrops and brought them back to Warsaw. This material included some well-preserved skull fragments belonging to pantodonts—primitive Paleocene mammals—from Naran Bulak in the Nemegt Valley, dinosaur eggs from the famous locality Bayn Dzak, and some individual dinosaur and mammal bones.

The winter of 1963/64 in the Paleontological Institute of the Polish Academy of Sciences was spent in preparing for the second expedition. Unfortunately, in the spring of 1964 Dr. Kulczycki fell seriously ill, and this ultimately kept him from participating. As for myself, I was just about to finish getting an extensive study on Paleozoic invertebrates ready for the printer, and could not be out of the country for the four months' duration of the expedition. The Institute accordingly asked Professor Kazimierz Kowalski, head of the Institute for Systematic Zoology in Cracow, to lead the 1964 expedition. I was slated to go out to Mongolia for a month, in July, to join in the work. Professor Kowalski, who is a zoologist and paleontologist as well as an enthusiastic traveler, readily accepted, but could not help with the work prior to departure because of a previous commitment at the Museum of Natural History in London in the Spring of 1964. I was accordingly ably assisted in the scientific preparation of the expedition by Dr. Andrzej Sulimski, an assistant at our Institute who had been on the reconnaissance expedition and was familiar with the Gobi's fossil fauna, and by two ladies who were to go on the 1964 expedition: Teresa Maryańska, an assistant of the Museum of Earth and Magdalena Borsuk-Białynicka, an assistant at the Paleozoological Institute of the University of Warsaw.

Two geologists invited to take part in the expedition prepared the cartographic material and studied the Gobi's geological structure. They were Dr. Ryszard Gradziński, of the Department of Geology of the Jagiellonian University of Cracow, and Dr. Jerzy Lefeld, of the Institute of Geological Sciences of the Polish Academy of Sciences in Warsaw. Maciej Kuczyński was again in charge of the technical organization.

In December 1963 the two Mongolian paleontologists from the reconnaissance expedition, Dovchin and Dashzeveg, arrived in Warsaw for their three months' training course. We immediately established friendly relations. They helped us in preparing the expedition's itinerary, taught us the Mongolian language, helped read Mongolian maps, and explained the transcription of Mongolian geographical names.

In March 1964 all the expedition's equipment was sent to Ulan-Bator by train, as had been done the previous year, and in mid-May we went to Warsaw airport and saw off the eleven participants of the second Polish-Mongolian paleontological expedition. On their way to Ulan-Bator they were to stop over in Moscow for a few days, to familiarize themselves with the Gobi collections on display at the Paleontological Museum there. The Polish group was expected to stay in Mongolia for over four months, including at least three months in the field. The first location was to be the Nemegt Valley, where excavations were planned in the Paleocene deposits at Naran Bulak, and in the Upper Cretaceous outcrops at Tsagan Khushu (also known as Tsagan Ula) and Nemegt. In August, during the second half of its stay, the expedition was to move to the Bayn Dzak outcrops dating from the lower part of the Upper Cretaceous, outcrops famous for their fossils of dinosaur eggs and primitive mammals. While the expedition headquarters was to be at Bayn Dzak, some of the members were to make a reconnaissance trip to the Trans-Altaian Gobi, which had not so far been explored by paleontologists.

5. Through the Gobi by Field Car

Lamasery on Gandan Hill in Ulan-Bator

Nemegt Valley tyrannosaur skeleton in the Municipal Museum of Ulan-Bator

Through the Gobi by Car

On my arrival in Ulan-Bator on June 13, 1964, I found that my team had already been camping in the Nemegt Valley in the southern Gobi for the past two weeks.

The sky was cloudless and the sun scorching on the day I reached the capital (population 200,000) of the Mongolian People's Republic. The air was very dry, much drier than that to I was accustomed in Poland, and for the first few days my throat was constantly parched. My first week in Mongolia was spent in visiting Ulan-Bator's relics of Mongolian civilization and its newly founded scientific centers. I became acquainted with the curriculum at the young (est. 1941) University of Ulan-Bator and with the work of the recently established research institutes of the Mongolian Academy of Sciences.

In the Municipal Museum of Ulan-Bator I was gazed at approvingly by the skeleton of a large carnivorous dinosaur—a tyrannosaur, genus *Tarbosaurus*—brought to light by the Soviet expedition to the Nemegt Valley. I also visited the palace, now a museum, of the last ruler of prerevolutionary Mongolia, bogdo-gegen Javsandamba IV, and the lamasery on Gandan Hill.

While in Ulan-Bator I picked up a number of letters my colleagues had left for me, containing much useful advice on how to prepare for the difficult journey and on how to reach their camp, including an accurately drawn diagram of the last part of the route through the Nemegt Valley.

At last the day of departure arrived. The Mongolian Academy of Sciences, whose attitude was extremely friendly throughout, had put at my disposal a small Gaz 69 field car with a Mongolian driver by the name Batochir. I was also to be accompanied by an employe of the Academy, Mr. Dagwa, who had spent a year and a half in Poland and spoke Polish well. Since the driver knew a little Russian, there was no communication problem. Early in the morning of June 19th we all climbed into the car and drove out of Ulan-Bator, due south over the vast, flat steppe. It soon became clear that Batochir had two points in his favor: he sang while he drove, chanting melodious Mongolian folk songs in a falsetto voice (seconded from the rear seat in a deeper pitch by Mr. Dagwa); and he was an excellent driver. In the steppe there are of course no roads in the proper sense of the word—only tracks made by the tires of passing cars. Several such tracks may run parallel, intersect, diverge, or join one another. Batochir had the uncanny gift of always finding the most convenient route. Occasionally he would take a shortcut through the steppe and make a new, shorter track. We drove fairly fast, 35 to 50 miles an hour—and even hit 60 where the ground was perfectly flat, and sped past grazing sheep and horses; countless steppe rodents ran in front of the car, and many found their deaths beneath its wheels.

Small rodents were not the only living things on the steppe. On a few occasions we managed to spot "tarbagan" whistlers (*Marmota sibirica*), whose smooth, brown, glossy fur is considered very valuable. Back in Poland we had been warned not to shoot tarbagans, since these rodents carry the plague; however, as our game was dinosaurs rather than tarbagans, and since we had been inoculated against plague, the danger was not too great.

While I had long known that the population density of Mongolia was $\frac{1}{150}$ that of Poland, it was only now, on this first journey south from Ulan-Bator, that I could visualize this fact for myself. Physically, it means that a population of one million (i.e., something less than the population of Warsaw) is spread out over an area five times the size of Poland. We were driving through a steppe practically without human habitations, save for lone nomads' tents and for a few larger settlements—"somon" seats—consisting of a few dozen tents and a few houses. Around noon we drove past an "obo" mound. These mounds, typical of Mongolian trails, are set up by travelers along the main caravan routes, each traveler being expected to add a few stones or some other object. The mounds thus keep growing over the years and may reach considerable size. Out of these vast, treeless, houseless steppes they form excellent landmarks.

We made a short stop at the obo, during which I prepared our lunch. We had a supply of canned food, bread, condensed milk, and fruit juice, plus a 3-gallon water jug. My companions ate only the meat and the bread, totally ignoring the fruit and vegetables. I recalled that the Mongolian paleontologists who had spent the winter with us in Poland had also refused to eat vegetables and had complained at the scantiness of the meat ration. Mongolians have the world's highest per capita meat consumption, and meat with milk forms their staple diet. They eat practically no fruit or vegetables.

After a short rest we were on our way again. We were scheduled to pass the first night in Mandal Gob, the capital of the Central Gobi aymak.* Just before reaching there we passed a large herd of camels of the Asian two-humped species *Camelus bactrianus*. Here the herds of horses we had kept passing while still near Ulan-Bator were much less numerous. The camels were standing in a cluster near a roadside well, waiting for a merciful traveler to water them. We stopped and poured a dozen bucketfuls of water from the well into the trough—but the camels drank the trough dry faster than we could fill it. We finally gave up when we remembered that a camel can drink 20 gallons of water at a time; it would have taken us hours to give them their fill.

The local authorities at Mandal Gob had been notified by the Mongolian Academy of Sciences of our impending arrival. We were

---
*An administrative unit corresponding to a province.

taken to a small hotel where rooms had been made ready for us. A Mongolian girl brought me a potful of hot tea, along with crisp cakes fried in sheep tallow (Mongolian tea is made from green, briqueted Chinese tea leaves boiled in milk, sometimes with sheep tallow added; it has the consistency and flavor of milk soup).

We resumed our journey very early the next day, heading for Dalan Dzadgad, our next stopover, and the capital of the South Gobi aymak. The sun blazed down mercilessly from early morning on. As we kept on southward, the steppe gradually changed to semidesert. Small, fast-moving lizards ran in front of our car in ever greater numbers, and the small rodents so numerous near Ulan-Bator were scarcer here. Camel herds lay tranquilly across the rutted car track without the least intention of getting out of the way, and would respond only to loud horn-tooting.

Since we had started early and had traveled at considerable speed, we reached Dalan Dzadgad at noontime instead of in the evening as planned. I was to get to know this town well during my frequent transits and longer stays in 1965. This was where we would usually store our rolling stock and our collections and where we would hire some of our laborers; here we were always treated with the greatest hospitality and friendliness and invariably were afforded all possible help. After dining as guest of the Darga* of the Dalan Dzadgad aymak, whom the Mongolian Academy of Sciences had notified of our arrival, I took a walk through town. Dalan Dzadgad has a summer population of about 5000; but in winter the arrival of nomadic shepherds swells this to 10000. There are a dozen or so streets of one- and two-story buildings in the center of town, but most of the population lives in yurts. The town's central spot is occupied by a fairly large public garden, which the authorities are determined at all costs to keep a green recreation center, in defiance of the surrounding desert, drought, hundred-degree summers and forty-below winters. The saplings and shrubbery could never have withstood the climate without the loving care they received; a system of pipes watered the garden day and night to keep the vegetation from drying out.

In Dalan Dzadgad I also picked up some letters left for me at the town hall by the other members of the team. I was to buy gauze to protect our faces from the swarms of tiny flies around the spring at Naran Bulak near camp. I was also asked to buy some bread, which was in short supply at camp. Having bought the items requested, we decided to push ahead the same day.

We drove due west out of Dalan Dzadgad. We had 25 miles to go to reach the pass in the Gurvan Saikhan mountain range. Once we reached this very picturesque pass, the air suddenly became cooler; our altitude was almost 7000 feet. We spent the night in

*Mongolian word for "chief."

Dismantling a yurt. Note the wooden frame

the village of Bayn Dalay, on the far side of the pass, and I slept in a *yurt* (tent) hotel for the first time in my life.

Every larger village in Mongolia has a hotel and restaurant for travelers. In the southern part of the country these are often yurt hotels. Having myself slept in such hotels, I can vouch for their comfort and testify that they remain cool on the hottest days; and I have been told that they also stay warm in the coldest weather.

A yurt is a large circular tent covered with felt and canvas. Yurts come in various sizes (ours was 15 to 20 feet in diameter). They are supported at the center by a tall, thick pole. The walls of the yurt, some 5 feet tall, are mounted on wooden frames. Rafters supporting the slanting roof extend from the top of the frame to the centerpole. The door leading in through the wooden frame is low, and has to be negotiated bending down; but away from the walls the yurt's interior is high and spacious. In the top of the roof, near the centerpole, is a hole that can be covered against the rain. The floor is carpeted with thick mats or camelhair rugs. Yurts in towns may have wooden floors. On hot days, the felt and canvas sides can be rolled up to expose the wooden frame, thus providing circulation and a relatively low temperature even during the worst heat waves. A hotel yurt holds six beds set up against the walls, with a washbasin to the left of the entrance. This consists of a bowl overhung by a vessel dripping water into it. Since water is in short supply in the desert, the strictest economy is imperative. We have found that with this kind of washbasin hands can be thoroughly washed in less than half a glass of running water, and that a quart of water is enough to wash the entire body. An iron stove (obviously for winter use only) stands in the center of the yurt, its flue pipe emerging through the upper hole; there are a few tables and chairs near the stove, and a small lavatory is provided under the wall, between the beds. In the somons of the southern Gobi these hotels usually consist of two yurts, and so can accommodate twelve travelers.

The yurts in which Mongol nomads live are somewhat smaller than those used in hotels. They are usually 12 feet in diameter, one yurt serving as home for an entire family. There is a small iron stove in the center (in olden days there was a fireplace). The yurt is divided into a women's section to the right side of the door and a men's section to the left; the former holds the food, vessels, and other household objects, the latter the equipment for managing the livestock. The yurts we had occasion to visit in the Gobi usually had a leather bottle with kumiss* hanging to the right of the entrance, anyone coming into the yurt being supposed to stir the kumiss once (to help the fermentation along). The place on the rear wall opposite the door is the family center. It usually contains

*Fermented camel's or mare's milk.

a cupboard or a wooden case with family photographs and souvenirs and often holds a radio receiver. If there is a party in the yurt, this is where the host and his most honored guests sit. There are usually two or three beds standing against the walls, sometimes with embroidered bed covers. Children usually sleep in beds, but in larger families the grownups sleep on rugs spread out on the floor.

Since nomadic Mongols change residence two, three, or sometimes even four times a year, the family cannot accumulate more belongings than can be easily transported. The traditional procedure is for Gobi herding families to move their belongings, including the yurts themselves, on camelback; nevertheless, we often encountered groups of Mongols moving their baggage by truck.

We had a very good night's sleep in the yurt, and I immediately felt at home in the morning when my breakfast tea was brought me in a Polish-made teapot just like the one I had back home in Warsaw, and when two Mongol boys drove past the hotel on a Czechoslovakian Jawa motorcycle. When the time came to leave, all the children gathered near the fence around the playing field in the center of the village to wave us good-bye.

Again we departed very early in the morning, to begin the most difficult part of our journey—to the somon of Gurvan Tes, the last village before our camp—the last 60 miles leading through a practically uninhabited waste of trackless desert and semidesert.

We were now in the Nemegt Valley, which runs about 110 miles from east to west and is 25 to 45 miles wide from north to south. This is a large, outletless depression typical of the Gobi region. To the north and south it is fringed by high massifs of Paleozoic and magmatic rocks, the valley between these massifs being a large tectonic trough filled with Upper Cretaceous and Lower Tertiary deposits.

The steep rocky slopes of the mountain ranges taper off into a smooth expanse of sandy Cretaceous deposits gently descending into the center of the Valley. After the infrequent but sometimes very abundant rains, water runs down from the mountains to the center of the Valley over the surfaces of these deposits, taking with it large amounts of gravel, sand, and mud.

The four massifs bordering the Nemegt Valley from the north are, from east to west, the Sevrey, Gilbent, Nemegt, and Altan Ula. The highest peak in the area, in the Nemegt range, has an altitude of over 9000 feet. The series of sandy Cretaceous deposits on the southern slopes of these high massifs have no individual names and are referred to as Nemegt, Altan Ula, etc.

The seasonal streams running down from the highest mountains usually exert a highly destructive force and cut deeply into the bedrock. This is why the Cretaceous deposits at the foot of the high

Haloxylon growth in Nemegt Valley. Sayrs in sandy Cretaceous deposits can be seen behind the bushes. The Nemegt massif is seen in the background

massifs are deeply scarred by numerous ravines (sayrs) which may extend—as they do at the foot of the Nemegt and Altan Ula massifs—over a dozen square miles. The water which periodically runs down these ravines carves out their sides, which are generally very steep and even vertical. The more strongly cemented sandstone banks—which occur amid Cretaceous and Tertiary sands and muds—form characteristic ledges on the sheer slopes. Since the ravines keep widening, isolated residual crests and separate hills are often formed. Rocky material from the surrounding mountain ranges is carried over the surface of the sandy Cretaceous deposits and forms a thin gravel layer over it. The winds, which never cease blowing, carry fine sand grains from the surface of these deposits to various places in the Valley, and this accumulates into sand dunes stretching for many miles.

The rainwater coming down from the mountains carries salt into the bottom of the Valley and forms brackish intermittent lakes in its center. One such lake was located near our destination of Gurvan Tes and was being worked for its salt.

Past Gurvan Tes three isolated camel-herder yurts were the only signs of human habitation. We drove over trackless semidesert ground with thinly scattered saksaul *Haloxylon* shrubs. The thick branches of this shrub, which is a typical feature of the Gobi flora, make excellent fuel and are stored by the local Mongols in summer as fuel for the winter (another precious fuel is argal, i.e., cattle dung). Otherwise there are no shrubs in the area except for a few species of Caragana bush: the thorny *Caragana spinosa*, whose branches are covered by fine golden bark, and the small-leaved *Caragana microphylla*.

It was now very hot and we often had to stop to prevent the engine from overheating. We were driving through a very unpleasant landscape of sand mounds overgrown with spreading *Nitraria sibirica* bushes. The fruits of this bush, sour berries that ripen in August, are eaten raw by the Mongols, who also stock them as food for the winter. I was particularly interested in this plant, since I knew that its roots carried a parasitic growth of goyo (*Cynomorium songaricum*), a plant typical of the Gobi. I had seen goyo only at the Botanical Institute of the Mongolian Academy of Sciences in Ulan-Bator, where its therapeutic properties were being studied. The stalk is edible. During one of our stops, I managed to find a goyo. Its stalk, sticking directly out of the sand, looked like a large cucumber covered with brown, velvety skin. Goyo usually grows in clusters, and what I found was one such cluster of brown "cucumbers." Later, in 1964 and in 1965, we were to encounter the goyo plant quite frequently. I took advantage of the stops to examine the different plants in the Nemegt Valley; even though the area is semi-arid, the flora is very variegated. Since I am not

45 Reconnaissance and Preparations

A typical obo mound on a Mongolian road

Goyo (*Cynomorium songaricum*). A parasitic plant indigenous to the Gobi

The Polish members of the 1964 expedition. Left to right—W. Skarżyński, M. Borsuk-Białynicka, R. Gradziński, J. Lefeld, A. Sulimski, G. Jakubowski, D. Walknowski, W. Maczek, T. Maryańska, K. Kowalski, and M. Kuczyński

a botanist, I could not identify most of the species, but I did recognize a few species of wormwood, *Salsola passerina,* and the driver Batochir drew my attention to the Gobi garlic, *Allium polyrhisum,* used by the Mongols in preparing mutton.

We had now passed the sand-mound region and were driving parallel to a picturesque range of barchans—dunes running along the Valley's east-west axis. Luckily, our way led along the dunes and not across them.

It was here that I first saw dzheiran gazelles (*Gazella subgutturosa*). A herd of seven of them was running over the hilly ground and across the path of our car. Our driver, his ancestral instincts seemingly aroused, accelerated to 50 mph in an attempt to cut them off. The animals were taken by surprise, enabling us to stop 20 yards or so in front of the frightened herd. They stood still only for a few seconds, but that was enough for me to note their lithe bodies, lyre-shaped, ringed horns, and large eyes. Batochir started the engine again and the herd began to flee. They now ran parallel to the track of our car at 40 miles per hour. After 15 minutes of this, they decided to turn off the road and flee. Though dzheiran herds are very frequently encountered in the open, semi-arid areas in the center of the valley, they appeared less often at our excavation sites, which were closer to the mountain ranges; instead, we saw mountain sheep (*Ovis ammon*) and ibexes (*Capra sibirica*).

Were it not for their cars' tire tracks on the gravelly top layer of desert sand, it would have been very hard for us to find the expedition's camp site among the endless hills and labyrinthine ravines. Near Naran Bulak, when we were at the point of losing our way, we suddenly saw a small signpost, "To Naran Bulak," made of two boards stuck into the ground. This was the work of our colleagues, who had foreseen the difficulties we would have in finding them. Finally, we saw a plume of smoke—they, too, had seen us from the camp, and had lit a smoke candle. Now we could find the place without any trouble. Thus I had finally arrived, after a two-and-a-half day crosscountry trip, bringing with me a supply of bread and of mail from Poland, both awaited with great impatience.

6. The Dinosaurs of Tsagan Khushu

Camp of Polish team at Naran Bulak in 1964, close to a dune overgrown with tamarisk shrubs

When I reached Naran Bulak I found that both of the Polish trucks had left for Ulan-Bator three days before, in order to bring in the remaining food, boards, plaster and other necessary items, since not all could be conveyed in a single trip. They were driven by our two drivers, Dobiesław Walknowski and Wiesław Maczek, with Maciej Kuczyński in charge. They had expected to meet us on the way, but their route was different—via Erdene Dalai—so we missed each other. The only vehicles now in camp were the Mongolian group's truck—the Gaz 63—and the small field car in which we had arrived. We therefore decided to use the field car to make a few reconnaissance trips to various points in the Nemegt Valley to help us plan future excavations. However, I wanted to spend my first few days looking at the work done so far.

The camp was divided into two groups, one working the Paleocene outcrops at Naran Bulak and the other working the Cretaceous formations five miles to the west at Tsagan Khushu.

The results of the first few days' work at Naran Bulak were very satisfactory. The area had been studied before by Soviet paleontologists, whose large-scale excavations had revealed a rich store of large fossil Lower Tertiary mammals—pantodonts and dinocerates. We therefore had expected no major discoveries. Contrary to expectations, however, we made a sensational find at Naran Bulak. Kazik* Kowalski, who specializes in small mammals, had found a thin layer—a lens, properly speaking—of very weakly cemented sandstone of near-sand consistency filled with fragments of small mammals hitherto undiscovered in the area. These fragments included lower jaws and skull parts of notoungulates (a primitive mammal group now extinct), as well as fragmentary remains of primitive lagomorphs, or insectivores, and of multituberculates. A similar mammalian fauna was found later during the 1964 expedition in the Tsagan Khushu Paleocene deposits by the Mongolian paleontologist Dashzeveg. Mammals belonging to these groups were already known to be present in the Gobi, but in a different area—in Khashaat, five miles from Bayn Dzak; this was the first time they had been discovered in the Nemegt Valley. The material collected served to complete the information available on the anatomy of these animals.

The Naran Bulak camp lay near a well; this greatly eased the problem of water supply, but also had its disadvantages. Clouds of tiny but very aggressive flies hung constantly over the well. Our reconnaissance expedition had made its visit the year before in July, toward the end of the season of the flies; thus, in 1964 we were totally unprepared for the insects. In June, a day after setting up camp

*Diminutive of Kazimierz.

at Naran Bulak, tired members of the expedition went bathing in a small stream issuing from the well; before they realized what was happening, they were bitten all over. A few even became seriously ill—swollen faces, blistered arms and legs, itching, and sleepless nights. Two days later the Polish group decided to move the camp a few hundred yards away from the well, to a point where the flies were much less annoying. The new camp site was at the foot of a small hill with a thick growth of saksaul *Haloxylon* and a few tamarisks (later we also encountered a single, tall tamarisk in a ravine while on a one-day reconnaissance trip to Altan Ula).

The flies, not the mammal fossils, were the main subject of conversation in camp, and were indeed our most persistent problem. The camp was a disaster area every day after sundown, which was when the flies came. Nobody dared go around in shorts or in a short-sleeved shirt. Heads were wrapped in towels and faces were protected by gauze. The sight of all of us sitting down to our evening meal rigged out this way had its lighter aspects. Nevertheless, our experts on mammals, who had to spend whole days lying on their faces picking small fragments of Paleocene mammals out of thin layers of sand, later maintained that of all the 1964 camps they enjoyed Naran Bulak most.

On the third day after my arrival I set out to visit the Tsagan Khushu group, leaving behind in Naran Bulak the mammal experts and our two geologists, who were finishing up the mapping of the area. The Polish group had only three people at Tsagan Khushu at the time: Teresa Maryańska, Gwidon Jakubowski, and Wojciech Skarżyński; but these three were accompanied by a group of laborers and were shortly to be joined by the Mongolian paleontologist Dovchin.

I had been told before leaving for Tsagan Khushu that I would pass through a little wooded grove on my way there. I suspected something was wrong, since I couldn't imagine trees growing in semi-arid country without surface water or topsoil. Nevertheless, we did in fact drive past a "grove" consisting of three lone Euphrates poplars growing out of a local pit where the water level must have been very close to the surface. At another spot between Naran Bulak and Tsagan Khushu we passed a single poplar tree, dead for many years. The dried-out trunk and the barkless, spreading branches only added to the desert gloom. These were the only trees we ever saw in the Nemegt Valley.

At Tsagan Khushu I was taken around by Gwidon and Teresa, who showed me the work done so far. The Cretaceous sandstone outcrops at Tsagan Khushu held layers full of very well preserved turtle shells. These included the complete shells of small land turtles about 8 inches long. We filled a whole case with them. There were also plenty of single dinosaur bones and pieces of fossilized tree trunk.

53 Dinosaurs of Tsagan Khushu

In the Nemegt Valley's Upper Cretaceous deposits, dinosaur bones are invariably found together with fossilized tree trunks, that grew there during the dinosaur era. The tree fragments were particularly numerous at Tsagan Khushu. There were petrified fragments of very heavy trunks about a foot in diameter, with the annual rings clearly visible in the cross section, and also smaller trunks and knotted branches. Unfortunately pollen, which could have served to provide a very accurate dating of the relative age of these layers, could not survive in the Valley's sandy deposits; so the age of its Cretaceous deposits had to be estimated on the basis of the dinosaur skeletons. This, however, is not as simple as it seems, since all the dinosaurs the Soviet paleontologists had found in this part of the world belonged to new species, i.e., to species not found earlier on other continents (including North America, which has the best stratification of the Upper Cretaceous continental series). And the dinosaurs which we have found in the Gobi belonged either to the species described by the Soviet paleontologists or to new species. This means that the geological strata in the Nemegt Valley must be worked out by comparing the assemblage of species appearing there with the assemblages of species known to appear in the Upper Cretaceous deposits of other continents.

Carnivorous species stood out among the dinosaurs found by the Soviet paleontologists in the area. Maleyev identified four different species, which he assigned to the two genera *Tyrannosaurus* and *Gorgosaurus* (each also known in North America) and to a new genus *Tarbosaurus*. This was later contested by Rozhdestvensky, who concluded that Maleyev's four species are conspecific, i.e., all belong to the single species *Tarbosaurus bataar*. The genus *Tarbosaurus*, which is typical of the Upper Cretaceous of Central Asia, corresponds to the North American genus *Tyrannosaurus*.

Another species found in the Nemegt Valley was a duck-billed dinosaur, identified by Rozhdestvensky as *Saurolophus angustirostris*. Genus *Saurolophus* was already known from the Upper Cretaceous of Canada, but the Canadian species differs from the Gobian. Finally, Maleyev found a new species of the armored dinosaur genus *Dyoplosaurus* in the Nemegt's Upper Cretaceous; this genus is also represented in the Upper Cretaceous of Canada.

Unfortunately, the various species of the genera *Tyrannosaurus*, *Saurolophus*, and *Dyoplosaurus* in North America's Upper Cretaceous formations appear not in one but in several horizons; it is therefore difficult to draw a parallel between the Asian and the North American beds. Nevertheless, analysis of the fauna in the Upper Cretaceous beds of the Nemegt Valley indicates that these beds correspond to the upper part of the Upper Cretaceous, i.e., to the Campanian or Maastrichian stages.

While I was walking with Gwidon Jakubowski and looking at the exposed light-colored and red Cretaceous sandstone strata, he showed me a place where bones were showing through a wall about 7 feet down. These bones were about 8 feet apart, and it could be assumed that they belonged to a single skeleton. We decided to start digging; if indeed a complete skeleton had been preserved it would not be hard to extract, since it was covered by only about 7 feet of loose sandstone.

After a day's digging we found our assumption to be correct: the bones visible in the wall were part of a skull, which we were able to expose almost undamaged and which proved to belong to a relatively small carnivorous dinosaur of the tyrannosaur group (probably to the common Gobi genus *Tarbosaurus*). On further exposure the skeleton became even more fascinating. After a few days we were able to gaze at an almost complete, excellently preserved tarbosaur skeleton lying on its side with its head thrown back, its legs drawn up, and its tail bent. The bones protruding from the other spot in the wall formed part of the tail, which was arched back with its tip close to the head. The skeleton found by Jakubowski was relatively small, perhaps 25 feet long. It had been preserved in the exact position in which the animal had met its death 80 million years ago. Dead camels are often found in the desert in the same posture of agony, head thrown back and legs drawn up.

The tyrannosaurs—the carnivorous dinosaurs of the second half of the Cretaceous belonging to family Tyrannosauridae—were the biggest and most dangerous predators of all time. Strangely enough, their forelimbs had shrunk so much that they were too short to reach the animals' mouths. Representatives of genus *Tarbosaurus,* whose skeletons are common in the Upper Cretaceous formations of the Gobi, reached (according to Rozhdestvensky) lengths of up to 46 feet and stood 20 to 23 feet tall. Tarbosaurs ran semi-erect on very strong, three-toed hind legs, using their long, muscular tails as props to maintain their balance. These giants had skulls some 4 feet long, and their strong jaws were equipped with very sharp and daggerlike serrated teeth up to 6 inches in length. Their forelimbs were very short, as in all other members of the family. Tarbosaurs preyed on the contemporary herbivorous dinosaurs, which they killed with their teeth and with their sharp-clawed hind legs.

The tarbosaur we uncovered at Tsagan Khushu was apparently young. Since it had lain at some depth (7 feet underground), its bones were unweathered and in an excellent state of preservation; only the last part of the tail (about 10 feet long), parts of the lower jaw, and the right forelimb were missing. We were very enthusiastic about our first dinosaur. When it was fully clear of the overlying rock we drew an exact copy of it and photographed it at least a dozen times from all possible angles (we found it very photogenic).

Our biggest problem was how to crate and transport this enormous reptile. The exposed skeleton weighed several tons including the rock enclosing it; in addition, it lay high up on a hard-to-reach rock ledge under which our trucks could not pass. We were therefore forced to cut the skeleton into sections for shipping. This proved to be far from simple. For example, in attempting to sever the head by cutting through the cervical vertebrae, we found that half the lower jaw had shifted and lay beneath the neck, so that we would damage the lower jaw by such a cut. We held long discussions on how to disassemble the skeleton with the least possible damage. Before cutting and crating we impregnated all cracks in the bones with thin polystyrene, which proved to be an excellent binder for bones. We also coated the bone surfaces with polystyrene in order to protect it from damage during transport. A special crate was built for each part of the skeleton, the first stage in crating being to construct a wooden frame tall enough to keep any of the bones from sticking out. The bones inside the framework were then covered with plastic foil and a thick layer of lignin, and plaster was poured into the empty spaces. The next stage was to detach the monolith from its base and turn it upside down. This was easily done once suitable incisions had been made all around the base. The crate was then inverted, the remaining holes plastered over, and the lid nailed down on the other side. The monolith was now ready to travel without danger of damage to the bones en route. The crating and transportation of our first dinosaur, which we named "Gwidon's skeleton," after its discoverer, took place in mid-July, after I had already left Tsagan Khushu. Back in Warsaw I received a letter at the end of July from Professor Kowalski saying that the packing and moving had gone very well and had taken sixteen crates, which were loaded on a truck and sent to Ulan-Bator before the end of the month.

This skeleton was not only the first dinosaur our expeditions discovered, but also the first ever to be prepared and mounted in Warsaw. Since it had been preserved in an exceptionally photogenic position and was fairly complete, we decided that before extracting all the bones from the rock it would be useful to make a cast of the skeleton in its original position. Therefore, after uncrating the monoliths and removing the protective plaster, we reassembled the lumps of bone-encasing rock in the exact position they had held at Tsagan Khushu. Before casting, some of the bones had to be partly freed from the surrounding rock, saturated with polystyrene, and have missing parts glued on. This work took a few months, after which the skeleton was ready for casting. This difficult operation was supervised by our Institute's technical assistant, Mr. Wojciech Skarżyński. The skeleton was divided once more into a dozen or so pieces and a negative plaster cast made of each. Once

Our first tarbosaur skeleton, Tsagan Khushu

57 Dinosaurs of Tsagan Khushu

Plaster cast of the tarbosaur skeleton

Excavation of the skeleton of a large ornithomimid at Tsagan Khusku. In foreground from left to right: T. Maryańska and W. Skarżynski. Standing: M. Kuczyński, R. Gradziński, Z. Kielan-Jaworowska

the large negative molds were ready, they were used to cast positives, which were reassembled into the shape of the skeleton. After they had set, the artistic work began. The cast had to be compared with the original, the blurred parts carved, the plaster bones painted their natural color, and the plaster in the spaces between the bones—corresponding to the enclosing rock—had to be covered with a layer of sand from the skeleton's old home in Tsagan Khushu. The final cast was an exact replica of the skeleton as we had originally seen it at Tsagan Khushu. We could now proceed with the preparation of the bones—the removal of the encasing rock and reassembly in the Museum.

Proud as we were of our first tarbosaur, we continued the search for new material in Tsagan Khushu. One day at dinner Wojtek* Skarżyński announced with an air of mystery that he had found something small but very interesting in the vicinity of Gwidon's skeleton. We proceeded to the spot immediately after dinner. He had in fact found a collection of fairly small bones from some small dinosaur. The bones of the hind legs, already exposed, were very well preserved and almost white. We took a closer look at them. These were relatively small, slender hind limbs with greatly elongated metatarsal bones, similarly structured as those of carnivorous dinosaurs. However, the skeleton could not have belonged to a carnivorous animal, since the proportions were different and the bones too fine. Outside of the flesh-eaters, we find this kind of hind-limb structure in only one other group of theropods—in the group resembling birds, particularly ostriches, in outward appearance. These are the ornithomimids found in Upper Cretaceous formations in Canada. The Soviet paleontologists had mentioned the presence of ornithomimid skeletal fragments in the Gobi, but had not given any details. Skarżyński was left behind to carry on the good work. He returned in the evening to announce that he had managed to find the skull. We found this hard to believe and, the late hour notwithstanding, went back to the site to inspect the newly exposed parts of the skeleton. Skarżyński had not been mistaken: there was the skull of an ornithomimid, the first ever found in Asia. Like the rest of the skeleton, it was relatively small, only about 6 inches long. It was somewhat flattened, but the bones of the cranial roof were very well preserved. The left side was encased in a hard lump of sandstone, but the exposed right side showed a large eye-socket, along with elongated, toothless jaws drawn out into a beak. Unfortunately, the cervical vertebrae and forelimbs had not been preserved. We had the skull, part of the pelvis, complete hind limbs, and an almost complete long tail. We calculated the animal's length to be about 8 feet, on the basis of the length of its hind limbs.

*Diminutive of Wojciech.

We were immensely excited by this new discovery, which overshadowed anything we had found up to that time. On our way back to camp we decided that our colleagues at Naran Bulak would have to be notified at once. Unfortunately, both of our trucks had stayed behind at Naran Bulak. Teresa Maryańska, the most stout-hearted of us all, offered to walk the five miles that separated us the very same night so as not to withhold the good news until the next day.

One morning we awoke to a dark, leaden sky and postponed our departure for work, expecting it to rain any minute. The rain did in fact come shortly afterward, but it was a peculiar kind of rain. The raindrops were clearly visible coming down, but in the dry desert air they dried up before reaching the ground, and only a few drops fell on the camp. It must have rained heavily somewhere nearby that night, however, since when we returned to dinner at the midday break Lefeld, who was working at another site, came back with the news that a river was flowing in the sayr. We all ran to see this for ourselves; a stream of coffee-colored water a dozen feet wide was in fact flowing in the wide, hitherto dry riverbed which our trucks had used as their road to Naran Bulak. It continued to flow for four hours, but the riverbed then dried out and hardened again, and the sayr could once more be used as a road.

This day brought us further discoveries. We had gotten into the habit of walking with a stoop, with our faces as close to the ground as possible, so as to keep an eye on the substratum. At Tsagan Khushu there were two horizons with large accumulations of bones; we referred to these as the upper and the lower ossiferous horizons, respectively. On the following day, while searching in the lower horizon, I chanced to move a block of sandstone that had detached itself from its foundation. The block contained the bent claws and phalanges of a forelimb. Part of a spinal column protruded from a rockface nearby. It was obvious that the bones belonged to the same individual, so it was probable that a large part of the skeleton had been preserved. We began extraction of this skeleton by digging a ditch all around it and then working our way in from the ditch, gradually removing the layers of sandstone and exposing the bones. Since the skeleton lay only slightly below the surface, it was strongly weathered, and some of the bones crumbled when touched. Teresa and I spent several days gluing and wrapping the disintegrating bones of the hind legs. When both the fore- and hind limbs were exposed, we found that this was yet another ornithomimid skeleton. However, unlike the one found by Skarżyński, "my" skeleton was large, at about 17 feet long one of the largest ornithomimid skeletons known. The animal was lying on its back, with its hind legs drawn up and its tail tucked in. The long, clawed prehensile forelimbs were thrown forward and were very well preserved. The pelvic bones were badly damaged, but all the tail vertebrae were perfectly preserved. The last, smallest vertebra was only $\frac{1}{5}$ of an inch long. We carefully numbered all the vertebrae and packed them in cardboard boxes. Unfortunately, as the front part of the skeleton was very imperfectly preserved, some ribs were missing and cervical vertebrae had disappeared without a trace. We assumed, however, that the ornithomimid's very long and supple neck was strongly bent backward (as is usually the case); this would mean that the skull, the most important part, would be found down beneath the pelvic bones. Now, the uncovering and removing of a medium-sized skeleton, including sketching the positions of all the bones, numbering them exactly, preparing the monoliths, and crating, usually took us about two weeks. Unfortunately, it came time for me to leave Tsagan Khushu before we had reached the pelvis. On my return to Warsaw I found a letter from Professor Kowalski stating that the skull had in fact been found under the pelvis, as expected.

Thus, the booty from our 1964 expedition included two ornithomimid skeletons, each with the skull intact. On our next expedition, in 1965, we made a specific search for ornithomimids, in the hope of finding at least one more skull. However, although we did manage to find a number of incomplete skeletons, numerous fore- and hind-limbs, a pelvic girdle, and fragments of spines, no more skulls of this group turned up.

7. Reconnaissance Trips to Altan Ula and Nemegt

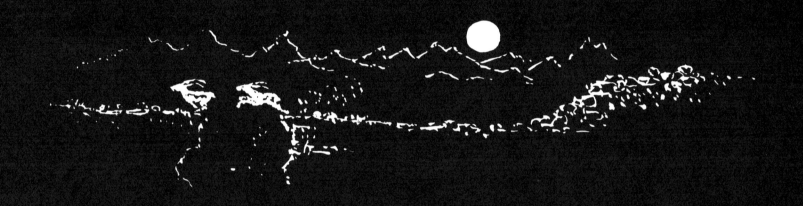

Ravines in Cretaceous sandstones of the Nemegt, seen from the east

During my stay at Tsagan Khushu in 1964 we used the field car for our reconnaissance trips to Nemegt and Altan Ula, the aim being to find the best sites for camping and digging. Since the car could only hold four persons besides the driver, the party included Dovchin (the head of the Mongolian group), Professor Kazimierz Kowalski, Dr. Ryszard Gradziński (a geologist with an excellent sense of direction in the field), and myself. The first problem was planning the reconnaissance. The obvious procedure was to make a three-day trip to inspect both the Nemegt and the Altan Ula outcrops. We knew, however, that both these places were waterless. The Soviet paleontologists who had worked there had had to bring their water from wells two dozen miles away. That meant we would have to carry enough water for three days. We assumed that working under desert conditions our average daily consumption of drinking water was more than a gallon per person; this brought the 3-day water supply needed for 5 people to over 15 gallons, even if we decided to dispense with washing during the trip. That was too much water, since a Gaz 69 with 5 passengers has no room for a tank of this size. We therefore decided to make two trips: a one-day trip to Altan Ula, and a two-day trip to Nemegt, farther away, on June 25–26.

The Altan Ula Cretaceous formations occupy four groups of sayrs very close to the Altan Ula Range. I have already mentioned that this is the westernmost range to the north of the Nemegt Valley.

Tsagan Khushu, which is located on the valley's central east-west axis, lies opposite the Altan Ula outcrops. It is impossible, however, to reach the Altan Ula sayrs directly from Tsagan Khushu, since the two places are separated by a dune range that is very difficult to cross.

The Soviet paleontologists had worked in outcrops which they had named Eastern Altan Ula and Central Altan Ula, and which we named Altan Ula I and Altan Ula II, respectively. The two series of outcrops farther to the west (Western Altan Ula), which we called Altan Ula III and Altan Ula IV, had not yet been explored by paleontologists because of their inaccessibility. We therefore decided to visit Altan Ula I and II first and then try to reach the outcrops farther to the west. We anticipated that the Altan Ula ravines could best be approached from the east.

We left camp at 6 A.M. Dovchin took a rifle, hoping that once near the mountains he would be able to bag one of the mountain sheep grazing in flocks on the mountainsides, or perhaps an ibex. On the way we expected to meet some of the dzheiran gazelles which frequently appear in small herds in the Nemegt Valley. Once again I admired the masterful way Batochir handled the car on the most difficult terrain. We crossed a dune range covered with a sparse growth of saksaul *Haloxylon,* which the Gaz 69 negotiated without any difficulty, and reached the fringe of the Eastern Altan

Ula outcrops quite early in the day. We agreed to work as follows: each new area we came to would be divided into four parts, with each of us exploring a certain group of ravines. We would meet back at the starting point three hours later. Dovchin was to explore the easternmost ravines, which ran down to where the vehicle was parked; Kowalski was to explore the group of ravines farther to the west; I myself was to inspect the ravines west of Kowalski's; and Gradziński was to work farthest to the west. This way we hoped to survey Altan Ula I and Altan Ula II in a relatively short time. At the beginning of our explorations the sun was still low in the sky and there was still some shade in the ravines. I was familiar with Rozhdestvensky's descriptions of the Altan Ula outcrops. This was where the Soviet expedition's driver Pronin had found the famous Dragon's Tomb 16 years earlier. These fragments of seven duck-billed dinosaurs discovered in a 7-foot-thick slab of very hard sandstone were subsequently identified by Rozhdestvensky as *Saurolophus angustirostris*. The question was how to find the Dragon's Tomb in this labyrinth of hills and ravines extending over a half-dozen square miles.

The ravine I was following had steep slopes, cutting into red sandstone with occasional intercalations of lighter-colored sand and gravel. I inspected the sides of the ravine, carefully turning over every loose piece of sandstone and sometimes climbing the walls to reach what I thought were bone fragments, but could not find a thing. The Cretaceous sandstone of Altan Ula appeared inhospitable and unfossiliferous. Suddenly my foot struck a piece of wooden board lying on the ground. I could scarcely believe my eyes. A board here? Clearly, it must have been left behind by the Soviet paleontologists. But this surely meant that their camp was somewhere nearby, and I was perhaps near the Dragon's Tomb. I began to inspect the surroundings with redoubled attention, kept circling along the next ravine and then along the next but one, and found more board fragments; yes, this was where the crating had taken place. At long last, I recognized the place familiar to me from the photographs: the traces of the Dragon's Tomb, slabs of hard red sandstone bearing the impressions of the ribs that had been taken out and sent to the laboratory. I wanted to shout for my companions, but this was not easy. I climbed the highest point close by and began calling out; 15 minutes later Kowalski arrived, and we inspected the Dragon's Tomb together. Technical difficulties had kept the Soviet paleontologists from extracting all the bones preserved in the hard sandstone. Unfortunately, we could do no better. The rock was too hard, the bones were in fragments, and the approach was too difficult. Neither in 1964 nor during the subsequent expedition did we find a sauroloph skeleton; this was strange since, judging by the experience of our Soviet predecessors,

A lizard of genus *Agama*

Skeleton of an argali (wild ram) in the Altan Ula mountains

sauroloph fossils are said to be common in the Nemegt Valley's Cretaceous formations. In addition to the incomplete Dragon's Tomb skeletons, the Soviet expeditions had uncovered a complete sauroloph skeleton in the northwest part of the Nemegt outcrops (this 25-foot-tall skeleton now stands in the lobby of the Moscow Paleontological Museum). On the other hand, even though we managed to find skeletons of dinosaurs hitherto unknown from the Gobi, we were unlucky as far as saurolophs were concerned, and failed to uncover any skeletal fragments aside from skull fragments with the typical set of teeth, skin imprints, and single bones.

We met Gradziński on our way back to the car. Dovchin had also returned. It was high noon and very hot. We ate lunch and then lay down under the car for a short rest, since this was the only shady place available.

At 2 P.M. we started off westward, toward the Altan Ula III outcrops, through enormous, tortuous sayrs. We separated again, each exploring a different ravine. On my lonely walk I failed to find any fossils but did pass an ibex skeleton lying on the ground. Large flocks of these goats inhabit the mountains of the Gobi, clambering about on the practically vertical slopes of the ravines with great agility. I picked up the skull, whose beautiful, tall, rearward-arching horns showed their annual growth rings very clearly; it was so heavy that I just managed to carry it the mile or so which separated me from our meeting point.

My companions, too, had failed to find any fossils, but this seemed to be our lucky day for present-day animals. Professor Kowalski had caught a beautiful black lizard about 12 inches long. The species, which belongs to genus *Agama*, occurs only in the Altan Ula part of the Nemegt Valley, so we had not seen it before. In Naran Bulak and Tsagan Khushu there is another kind of lizard (*Phrynocephalus versicolor*), which, unlike the Altan Ula lizard, is small and very nimble, with a small arched tail. The captured lizard was put into a pouch, to be taken to Naran Bulak and to await the arrival of Maciej Kuczyński, who was making a film on the flora and fauna of the Gobi. It reached its destination safe and sound and was kept in a box for a few days; but Maciej was not destined to film it, since it escaped before his return.

It was only on our way back to the car that we noticed some dinosaur bones at two separate points on the rock slope. It was difficult to secure any skeletons on so short a stay, but we realized that these strata held major accumulations of fossils; later, during our 1965 expedition, we had the opportunity to explore the riches in this area.

Two days later we made a two-day reconnaissance to Nemegt. The team was the same as before. We approached the Nemegt cliffs from the east, at first encountering an unfossiliferous sandstone series. Then, as I was walking along one of the most picturesque ravines in the Nemegt, I suddenly heard the noise of falling stones quite close to me. My immediate reaction was alarm at an impending avalanche of overhanging sandstone, but I soon saw the real reason for the incident. A beautiful mountain ram—an argali—was standing right in front of me, no more than 15 feet away. Argali are among the largest of wild goats; they sometimes stand $4\frac{1}{2}$ feet high at the shoulder, with the head $5\frac{1}{2}$ feet off the ground. This was the size of my animal. I managed to note the mighty horns—touching at the base and diverging at the top with a twist-and-a-half, the gray-white fleece, and the broad white beard under the throat. Unfortunately I had no camera with me, and probably couldn't have used one anyway, since the frightened animal took flight immediately. Never again in my long stay on the Gobi in 1965 was I so close to an argali, even though in 1965 I watched whole flocks of them through a field glass in the Altan Ula hills.

The excavation-site locations reported by the Soviet expedition were none too accurate, and it was only after several hours that we realized that the bone-bearing beds from which they had removed their dinosaur skeletons lay a few miles to the west of the area we were reconnoitering. Only late in the afternoon did we reach sayrs with the lighter-colored, yellowish sandstone rich in fossils. Since the Soviets had marked most of their campsites on their maps, we kept looking for traces of these sites in order to get an idea of the approximate locations of the excavation sites themselves. However, we had a very large area to cover—several square miles of outcrops—and it was far from easy to find traces of a campsite in such a labyrinth of tortuous ravines. Toward evening, when we were just about ready to give up, I noticed, in the light of the glancing rays of the setting sun, some very faint parallel impressions in the gravel layer covering the desert sand. These were undoubtedly tire tracks that had become pressed into the gravel layer and had survived for 15 years. We followed them for two miles and reached what had been our Soviet colleagues' camp site: peg holes, a few empty food cans, pieces of wood, even a broken-off heel. Now that we had located the campsite we could finally orient the Soviet workers' sketches. Our first few hours of searching in this area were promptly rewarded with finds of single bones, such as phalanges and vertebrae, and fragments of other bones. It was obvious that the beds were rich and that major accumulations of fossils were probably present.

Both in Altan Ula and in the Nemegt the Soviet workers had left their mark, "Paleontological Expedition of the Academy of Sciences of the USSR, 1948," carved into the sandstone in Russian; these marks had survived for sixteen years.

It was now July, and the time approached for me to leave the

Jerboa (*Allactaga bullata*) from the Nemegt Valley

Nemegt Valley. A day before I was due to depart, the two Polish trucks returned to camp from Ulan-Bator with fresh supplies of food, plaster, and wooden boards.

I rode back to Ulan-Bator by a shorter route, which lead through Bayn Dzak and bypassed Dalan Dzadgad, accompanied by Dagwa, who had greatly enjoyed his stay on the Gobi, and by Batochir. As we were leaving the Nemegt Valley at nightfall, I saw the jerboa (*Allactaga bullata*) for the first time. This is a family of small (length about 8 inches not including the tail), very shapely rodents, with large heads, and very long legs and tails. Their large eyes indicate that they are night animals—they remain in their lairs throughout the day and go out in search of food only after dark. They are very fast runners, leaping high into the air with their strong legs while keeping their balance with the aid of their tails. Even though they are very numerous in semi-arid areas, they are seen only infrequently, thanks to their gray-yellow protective coloring; they are also very shy and disappear into their lairs at the slightest noise.

On my return to Ulan-Bator I paid a visit to Professor Shyrendyb, Chairman of the Mongolian Academy of Sciences, and Mongolian sponsor of our expeditions, who always showed much benevolent interest in our work. We discussed the work of our projected third and last expedition, scheduled for 1965. The knowledge that I would return for a longer stay the following year made it easier for me to leave sunny Mongolia.

8. The Trans-Altaian Gobi

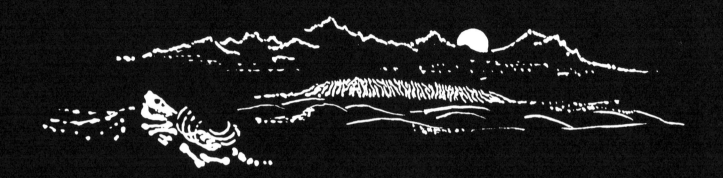

Skeleton of a small tarbosaur extracted in 1964 in the Nemegt.

On my departure from Tsagan Khushu, we agreed with Professor Kowalski and Dovchin that the work at Naran Bulak would be terminated soon afterward and that everybody would move to Tsagan Khushu to help with the difficult job of crating and transporting the two large skeletons.

On the 4th of July a group of five persons—Gradzínski, Jakubowski, Kuczyński, Maryańska, and Walknowski—drove out in the Star 66 for a three-day reconnaissance of the westernmost Altan Ula outcrops, which we had named Altan Ula IV and which we had not been able to reach on our June trip. This July trip was of major importance for our subsequent work, since it turned up a track by which to approach the Altan Ula IV outcrops; this track was very rough, leading as it did through a dune range several miles long, but our Star 66 was able to negotiate it. This brief trip to Altan Ula showed that these sandy Upper Cretaceous deposits were very rich in dinosaur fragments. On the very first day, Walknowski found a well-preserved tarbosaur skull lying at a shallow depth, along with the large, macelike tail of a carnivorous dinosaur belonging to genus *Dyoplosaurus*. They also found a sauropod shoulder blade, a dinosaur egg, and several isolated bones of various reptiles. Since the group had promised to be back at Tsagan Khushu on the appointed date, and in consideration of the difficult road conditions and lack of contact with the base, not all the specimens could be uncovered and taken back. It was therefore necessary to make another trip to Altan Ula. In the meantime the work at Tsagan Khushu had been all but completed. July 10th and 11th were the Mongolian national holidays—Nadom—celebrated by all Mongols with great ceremony. Dovchin decided to take all the Mongolian members of the expedition to the Nadom celebrations in Dalan Dzadgad; the laborers, who had completed their work, also left. One of the Polish cars, the Star 25, left for Ulan-Bator, with the skeleton of the Tsagan Khushu tarbosaur on board; on its way back it was to bring a fresh supply of plaster and boards from Ulan-Bator. The Tsagan Khushu camp was left practically deserted, only a few members of the Polish group remaining; these proceeded to make a second trip to Altan Ula in the Star 66. Ryszard Gradziński was left alone to hold the fort. The work at Altan Ula was successful, which was more than could be said about the camp at Tsagan Khushu. On July 11th it was visited by a hurricane-like sandstorm. Not a single tent was left standing, the awnings over the tables were torn, and countless objects were carried away by the wind. Gradziński was only able to save the tents from utter destruction by weighting them down with heavy objects. By the time the Star 66 had come back from Altan Ula with the specimens, the camp looked like a battlefield.

On July 12th camp was struck at Altan Ula, and the expedition

Trans-Altaian Gobi. Yurts in the Bayn Gob somon

A poplar oasis in the somon of Zakhuy 77 The Trans-Altaian Gobi

moved to Nemegt for a three weeks' stay.

In this short period one small tarbosaur skeleton, only 13 feet long, was uncovered. Found by Magdalena Borsuk, it became of course "Magdalena's skeleton." This practice of naming skeletons after their discoverers was very convenient. If, say, somebody found two tarbosaur skeletons in the same spot, they were referred to, respectively, as Wojtek's first and Wojtek's second tarbosaur, making everything clear. To the uninitiated, however, this language might have sounded strange: in 1965, for example, we would speak of Dovchin's vertebral column, Małecki's jaw and leg (there bones could not be given a first name, since Dr. Małecki's first name is Jerzy and there was another Jerzy on the expedition), Teresa's and Zofia's skeletons, Wojtek's tail, Barsbold's pelvis, etc.

Magdalena's skeleton was not large, but made up for it by being very photogenic. It was lying on its side, with its head thrown back and its arched tail describing a circle reaching almost to the back of its head. Many skeletal fragments of other reptiles were also uncovered.

Toward the end of July the expedition moved from Nemegt to Bayn Dzak. Bayn Dzak will be described in one of the following chapters; for the present I shall merely mention that this place, discovered by an American expedition and famous for its dinosaur eggs and skulls of primitive mammals, could legitimately be expected to yield valuable finds. Our group got there on July, 30th, and by the next day Professor Kowalski had already found the first Cretaceous multituberculate skull. On August 4 five Poles and two Mongols left aboard two automobiles for a three-week reconnaissance of paleontological virgin soil in the Trans-Altaian Gobi. The personnel left behind in Bayn Dzak, with Andrzej Sulimski and Dashzeveg in charge, were supposed to keep looking for mammal skulls. The results of the five weeks' stay in Bayn Dzak were very satisfactory. Finds included nine skulls of rare Cretaceous mammals—insectivores and multituberculates—a few skulls of small dinosaurs, which proved to be *Protoceratops* (the ancestor of the horned dinosaurs), and numerous dinosaur eggs. One of the most valuable specimens was the almost complete skeleton, including the very well preserved skull, of a small armored reptile of genus *Pinacosaurus,* hitherto known to have existed only on the strength of one incomplete skull found by American expeditions at Bayn Dzak.

Back in Warsaw, I was kept informed on the progress of the work in Bayn Dzak during August by means of letters, which kept arriving fairly regularly during that period. Bayn Dzak was only ten miles away from the village of Bulgan, which had a post office, so letters could be sent without difficulty. On the other hand, there was no news for a long time from the Trans-Altaian Gobi group, since communications with this area are very difficult. At long last I received an extensive report from Professor Kowalski who wrote as follows:

"On August 4th we headed due west out of Bayn Dzak on board the Star 66 and the Mongolian car. There were five people in the Polish group: Teresa Maryańska, Maciej Kuczyński, Ryszard Gradziński, Dobiesław Walknowski, and myself; Dovchin and the driver Khorloo rode in the Mongolian car. We passed Bulgan and turned east. A short distance past Bulgan we entered a desert area literally covered with chalcedony (each of us made a collection for himself and had much trouble deciding which stones to keep, they were all so beautiful). We drove on through the steppe, and for our first and only time in Mongolia encountered a flock of vultures—enormous, ominous-looking birds. Once through the pass in the Atch Bogdo range we were back in the desert. The midday rest period was spent near sand dunes literally covered with flints. We inspected a few outcrops of red sandstone on the way, but found no fossils.

"On the following day we could see the Nemegt and Altan Ula ranges on the horizon. We stopped for the night at the foot of the Ikhe Bogdo mountains, which still had snow clinging to their tops in isolated spots. We continued on our way through Bayn Gol, where we met a group of Mongolian geologists which included Barsbold, who was well known to the members of the first expedition. We passed the night in the mountains at an altitude of over 6000 feet. The following day we reached a large desert valley to the north of the Somon Khayrkhan range, whose tallest peak is a fantastic, inaccessible rocky spire. According to the map, this range can be crossed to reach the valley to the south of it. We did in fact reach the top of the range, after a nightmarish ride through sayrs and dunes, but the granite slopes on the south side turned out to be vertical and could not be negotiated. We spent the night in the mountains and returned north the next day. The trip was not entirely wasted, however, since on our way back we spotted red-colored outcrops and went to inspect them (after chasing away a herd of grazing dzheiran). These outcrops proved to be full of vertebrate fragments. We found an assemblage of small Oligocene mammals and a large number of big bones, including a rhinoceros jaw. The cliffs are called Ulan Ganga, i.e., Red Valley.

"Toward evening we stopped collecting and drove toward the west, through enormous prairies and salt marshes. The country was very beautiful, full of clusters of flowering tamarisks, but swarmed with gnats and mosquitoes. To the northwest we found more large outcrops containing Oligocene mammal fossils. We passed the night on the spot, and continued our searches the following morning. Pushing on westward, we reached the somon of Zakhuy by evening.

Granite massif of Khatan Kyayrkhan, seen from the somon of Zakhuy

The somon of Khalyun, Trans-Altaian Gobi

81 The Trans-Altaian Gobi

While still some distance away, we were astonished to see a forest, something quite unusual in the desert. There stood a real cluster of poplars on the fringe of prairie and salt marshes, with the houses of the somon standing among the trees and an actual small river flowing nearby. We talked to the Darga of the somon: the district is larger than Belgium and Holland combined but has only 1500 inhabitants, along with 50,000 sheep and camels. Most of it is totally waterless Trans-Altaian Gobi.

"In the morning we picked up a local guide and drove the Star 66 out to the south of the Edrengyeen Nuru mountains. We crossed the range via a rocky canyon in the fantastically colored metamorphic rock. The area stretching south from these mountains to the Chinese border is quite uninhabited and waterless, a desert overgrown with saksaul *Haloxylon*, most of which is dead, probably because of a particularly long-lasting drought. This is the closest approach to an absolutely barren desert in Mongolia, with practically no plants, and with none of the rodents or insects that abound in other parts of the Gobi. We searched a few reddish outcrops but came up with nothing. We then climbed a high mountain of black volcanic rock covered with a desert glaze, and got a panoramic view to the south. The view was, for a paleontologist, quite discouraging, since there were no extensive outcrops and thus no potential discoveries. After a night around a fire we returned to Zakhuy; near the village we found a small layer of white sand with unidentifiable fragments of mammal bones and turtle shells. In the evening the somon authorities gave a festive reception in our honor. The Mongols sang their truly beautiful songs and demanded that we reciprocate. It turned out that singing was not exactly our forte, and the only songs we could execute passably in chorus were Christmas carols. When requested to translate the words, we got out of the difficulty by claiming they were old revolutionary songs.

"On the following morning we set out westward. The Zakhuy oasis is fringed by the beautiful granite mountains of Khatan Khayrkhan. The sand-carrying desert winds have polished them smooth, and the passes are covered by huge sand dunes which from a distance look like glaciers. They are probably the most beautiful mountains I have ever seen. The foothills showed promising red-colored outcrops but were at first difficult to reach, since they were separated from us by clay fields partly sown with barley and criss-crossed by irrigation ditches. Worse still, we found we could not turn back either, since we would then be driving with the wind, and the dust raised by the wheels would reduce visibility to practically zero. We decided to pass the night at the foot of the mountains. On the smooth granite blocks we saw some large, black *Agama* lizards that looked like small dinosaurs.

"In the morning we finally managed to find our way to the red outcrops, which again proved very rich in Oligocene bones. We named the site Khatan Khayrkhan, after the mountains. In the afternoon we were again on our way, in very high spirits, since our finds just about gave us the age (hitherto unknown) of the deposits filling the Trans-Altaian depressions. We soon encountered a beautiful prairie spring, in which we bathed with great joy, washing off the dust of the past few days.

"On August 14th we set out in the Mongolian truck for a summer yurt settlement situated on a 10,000-foot-high plateau. The nearby hollows abounded in whistlers and marmots. We spent the night in a yurt. The next morning we found horses waiting for us on the hoarfrost-covered grass to take us up to the summit. After a very tiring ride (we were poor horsemen, and had to use the uncomfortable Mongolian saddles in the bargain) we reached the summit, 12,553 feet above sea level. There was a huge obo, and the breath-taking view extended far beyond the Chinese border. Spring flowers were blooming under blankets of snow, and the meadows were covered with an abundant growth of edelweiss. We returned to camp in the evening, bringing back not only the members of the expedition, a large number of Mongols and a folded yurt, but also a live ram for our supper.

"On the following day, soon after departure, we found extensive outcrops at an altitude of over 6000 feet, where a seasonal stream keeps scouring a large trough of its contents of Oligocene clays, sands and gravels. In addition to numerous small mammals, we found the giant bones of the hornless rhinoceros *Indricotherium*. We continued westward the same evening, driving on through the moonlit night, and reached a ravine, one side of which consisted of metamorphic rock and the other of sheer cliffs of Oligocene clays. Its bottom had a fairly abundant plant cover (grass and flowering tamarisks), thanks to the Khaitch springs below. The ravine also entombed a fair amount of Oligocene fauna.

"Two days later, in fog and rain, we crossed the main Altai range and a deep valley and entered a little valley to the north, in the somon of Khalyun. The northern slopes of this little valley showed red Oligocene outcrops about 40 miles long, in which we found bones and teeth belonging to rodents of the family Cylindrodontidae. We then drove to the somon seat to change an axle spring which had broken on the journey. On August 20th we reached Yessen Bulak, the capital of the aymak (today the place is called Altai, but the old name still appears on the map). The township has a fairly large museum, with interesting ethnographic specimens and a few bones of Oligocene Cylindrodontidae. We rode east out of Yessen Bulak along the main road to Ulan-Bator; the road passed through beautiful mountain meadows and was very picturesque,

Oligocene outcrops at Khaitch, Trans-Altaian Gobi

but without interest for a paleontologist. On August 25 we turned off to the south and came to Tatal Gol—an Oligocene site visited first by American and then by Soviet expeditions. There were several parallel vales cutting through Oligocene clays intercalated with a layer of lava, with very numerous bones (mainly rodent) washed out to the surface by the rains; we managed to collect hundreds of these in a couple of days. We also looked for bones in the overlying strata, which are known as the Miocene Loh formation, but only unidentifiable fragments could be found. We then made a sortie northward along a sayr from Tatal Gol to a place known as Ondai Sayr, whose shales contain numerous fragments of Cretaceous plants and invertebrates. Fish skeletons were also encountered.

"We started back on August 28th, and reached the Bayn Dzak camp the following day; we were then told of the magnificent discoveries of Cretaceous mammals and were given letters from home..."

I read Professor Kowalski's report with great satisfaction, since it pointed out a number of new sites with Oligocene mammal faunas in the Trans-Altai Gobi.

With the Trans-Altaian group's return to Bayn Dzak, field work was brought to a close, and on September 10, after one hundred days in the field, the expedition arrived in Ulan-Bator.

During September the Polish members of the expedition returned to Warsaw, in three separate groups. The last, made up of Professor Kowalski, Mr. Kuczyński, and Wiesław Maczek, only got back on September 29th, after final liquidation of the work and shipment of the collections to Poland. At last we were able to thank Professor Kowalski for his able leadership and for the encouraging results obtained.

9. Getting Ready for the 1965 Expedition

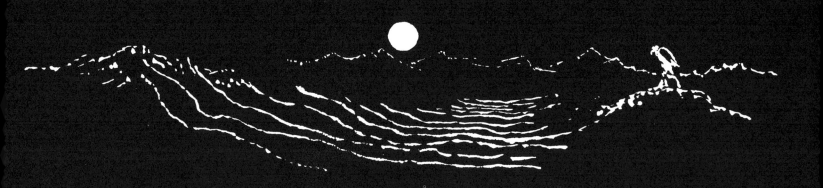

Halt under a lone desert elm on the way to Dalan Dzadgad

We passed the winter of 1964/65 in preparing for the new expedition. The 1964 expedition's entire outfit had remained in Ulan-Bator, stored in a large shed in the yard of the Academy of Sciences. The two Polish trucks were also left in Ulan-Bator for the winter.

On our return to Poland we asked the Starachowice factory for the loan of another six-wheel-drive Star 66, and the management kindly consented. As they were interested in the Star 66's performance under desert conditions, their motor mechanic, Mr. Edmund Rachtan, was to join our expedition; he was to prove to be an indispensable companion.

The list of participants had changed somewhat. Professor Kowalski was about to spend a few months in the United States and was unable to take part; two persons had to beg off in order to complete their theses, and two others for personal reasons.

The most experienced Gobi veterans, i.e., those from the first two expeditions who were now going to Mongolia for the third time, were Maciej Kuczyński and Dobiesław Walknowski. Five people—Teresa Maryańska and Wojciech Skarżyński, our two geologists, Ryszard Gradziński and Jerzy Lefeld, and myself—were going for the second time. Of the fifteen people on the 1965 expedition, eight were inexperienced and had never been to the desert before.

The 1964 expedition had had no cook. Meals had been prepared by kitchen orderlies in daily rotation. Breakfast had invariably been prepared (at his own request) by Jerzy Lefeld, who had specialized in porridge-making and was excused from preparing dinner and supper. The kitchen orderly would go out into the field in the morning with everybody else, but would have to return in time to make the fire and prepare dinner. In accordance with desert practice, we used saksaul *Haloxylon* as fuel; it had the advantages of being dry, of burning brightly, and of giving abundant heat. While the others were resting after lunch, the kitchen orderly would do the dishes. Owing to the need to economize water, we had not been able to adopt the obvious course of everyone washing his own mess kit. In the afternoon, the kitchen orderly would also have to cut work short and go back to prepare supper. This meant that one person was always unavailable for work, and since the camp was often divided into two groups, the proportion of personnel unavailable for excavation work was even higher. We accordingly decided that the 1965 expedition would have a cook. Cooking for 15 persons in a field kitchen at 104° in the shade was going to be hard work, especially since the cook would also have to boil a few dozen gallons of drinking water every day, as unboiled water could not be drunk. I was also of the opinion that a group of 15 persons working in the desert 60 miles away from the nearest human habitation and 250 miles away from the nearest first-aid station should also have a physician. In 1964 we had had no physician,

Ryszard Gradziński acting as medic, but this time there would be more of us and we would be working at more remote sites. Clearly, since the physician to a group of healthy individuals would be unemployed most of the time . . . he could take on the duties of cook as well. We accordingly decided to insert the following ad in the newspaper "Życie Warszawy": "Scientific expedition to the Gobi Desert will engage medical officer prepared to act as cook for 15 persons under field conditions. Departure in summer for 4 months' stay. Apply to Palaeozoological Institute of Polish Academy of Sciences."

Within a few days we had managed to interview two dozen young physicians willing to take part in the expedition and do the cooking and washing up. Our final choice was a young surgeon named Marek Łepkowski, then a junior physician at the Tuberculosis Institute. Luckily enough, there was no call for his professional services during the expedition, but he proved to be an excellent and very conscientious cook, and, best of all, a very pleasant companion.

We were also short one driver. This time we were to have three Polish trucks; one Star 25 and two Star 66's. Wiesław Maczek, who had driven the Star 25 in 1964, could not go on the 1965 expedition. This was especially regretted, since he had proved himself a jack-of-all-trades; he could both drive and repair a truck, was a good cook, a good fossil hunter and, in addition, a real game hunter, who at Bayn Dzak would often bring back a hare for dinner.

Since in 1965 we were to have Mr. Rachtan, the professional motor mechanic from the Starachowice Factory, and also Mr. Walknowski, an artist and a driver-mechanic by inclination (who was moreover going out to the Gobi for the third time and was therefore familiar both with our Star trucks and with their performance in the desert), we decided that we could afford to take on an amateur as our third driver. Our Star 25 was thus to be driven by a junior member of the expedition, Mr. Józef Kaźmierczak, who worked at the Institute as an assistant; he had been through a training course in an automobile workshop and was an excellent driver and repairman.

Even though all the 1964 expedition's equipment had been left for the winter in Ulan-Bator, we still had to buy many items in Warsaw, including gasoline, wooden boards, plaster, glue, clothing for the new members, a first-aid kit, food, etc. As usual, Mr. Kuczyński was in charge of these activities. The expedition's ladies—this year there were to be three: Teresa Maryańska, Dr. Halszka Osmólska (a scientific assistant at the Institute), and myself—decided to take a closer look at the food problem. The 1964 expedition had had a great variety of canned meat, vegetables, fruits and preserves, but no condiments to improve the insipid flavor of the canned food. We therefore drew up a list of various additives and condiments such as bay leaves, marjoram, cinnamon, vanilla, citric acid, yeast, gelatin, etc., and handed it to Mr. Kuczyński. We had not written down the amounts to be bought, assuming that Mr. Kuczyński, an expert in calculating foodstocks in terms of calories per person per day, would know how much of each item to buy. As he was in a hurry, however, he merely gave the list to two young students who were employed to do our shopping for us. These gentlemen had been accustomed to buying all our provisions in bulk. Therefore when the supplies were opened in the field, we found that we were in possession of a whole case of condiments, something which would undoubtedly have been worth a small fortune in the days of Columbus. These supplies included, among other things, four pounds of dried yeast, enough to supply a dozen expeditions such as ours with yeast cakes every evening; this was presented as a gift to the Dalan Dzadgad bakery at the end of the expedition.

During the winter we once more played host to visitors from Ulan-Bator. Dashzeveg, who was about to publish his second paper on Oligocene mammals of the Gobi in our quarterly "Acta Palaeontologica Polonica," arrived for a stay of two months; we also had a young Mongolian laboratory assistant, Erdenibulgan, for six months of training in the preparation of fossils. In our agreement with the Mongolian Academy of Sciences we had undertaken to donate all the equipment for the newly founded Ulan-Bator Paleontological Laboratory; so we were busy buying the necessary equipment (binoculars, microscopes, preparation apparatus, photographic material, etc.), while Erdenibulgan was learning how to use it.

We were also busy preparing the expedition's work program. The Upper Cretaceous formations in the Nemegt Valley had been explored by the Soviet expeditions and subsequently by ourselves in 1964. The main work had been done in the Nemegt, Tsagan Khushu, Eastern Altan Ula and Central Altan Ula outcrops. On the other hand, the westernmost Altan Ula outcrops, which groups of our paleontologists had visited twice in 1964 on reconnaissance trips from Tsagan Khushu, remained practically unexplored. The reason was clear. The only way to get to Western Altan Ula is over a range of dunes several miles across, in the center of the Nemegt Valley. This difficult approach had so far discouraged all attempts at exploration. We knew, however, that the dune range had been crossed twice in 1964 by our Star 66, so we could reasonably make another such attempt in 1965, when two such vehicles would be at our disposal.

According to the work program the team was to be divided into two groups, each working in a different place. The larger group was to set up camp in the Nemegt Valley and to work at Western

Altan Ula, Tsagan Khushu, and Nemegt. The smaller group was to work in Bayn Dzak during the first half of the period and to concentrate on Cretaceous mammals. The two groups were then to meet and form another detachment to go to Western Mongolia, where very extensive Upper Tertiary (Pliocene) deposits with mammal faunas had been discovered in 1949 by the last Soviet expedition.

A delegation from the Board of the Polish Academy of Sciences was leaving for Ulan-Bator in 1965 to extend the scientific collaboration agreement between the two Academies for another two years. I was attached to this delegation and left Warsaw by air on May 13th. The other members of the expedition were to reach Ulan-Bator in two groups during the second half of May. All the equipment for the expedition had been sent by rail to Ulan-Bator in March in accordance with the usual practice, and was awaiting us in Ulan-Bator.

We spent two days in conversations with the authorities of the Mongolian Academy of Sciences. The expedition's program for 1965 was approved, with insignificant changes; we were requested to cut down our stay in the field somewhat, since our Mongolian colleagues had to be back in Ulan-Bator before the end of August. It was also feared that we might have difficulty entering the western part of the country, since the area had been quarantined on account of a foot-and-mouth disease epidemic.

The technicians' group got to Ulan-Bator on May 19th, and the remaining members of the expedition arrived on May 24th. We intended to leave on June 1, so there was a lot of work to do. Every morning we put on our overalls at the hotel and went out to work in the storage shed. The three Star trucks were standing in the courtyard of the Academy, ready to be loaded. We also hired another truck—a 10-ton Tatra with a Mongolian driver—to carry a large part of our equipment to the Nemegt Valley. The Star 25, which was to go to Bayn Dzak, was to carry all the material for the group which was to work there. The Mongolian group now had two Soviet-made Gaz 63 trucks at its disposal; one of these was to go to Bayn Dzak and the other to the Nemegt Valley. Dovchin was to head the Mongolian group in the Nemegt Valley, while Dashzeveg and Erdenibulgan were to go to Bayn Dzak. We also hired a small Gaz 69 field car, which we used for short reconnaissance trips during our first three weeks in the Nemegt.

By June 1st our truck convoy was loaded to capacity, and we drove off. We were all going together as far as Dalan Dzadgad, and there we were to split up into two groups. We arrived there, exhausted, after $2\frac{1}{2}$ days of travel. The familiar journey from Ulan-Bator to Dalan Dzadgad was uneventful, and we spent the nights out in the open in our sleeping bags. It was already very hot in the daytime, but the first night was frosty and we found our sleeping bags covered with hoarfrost in the morning.

Our itinerary now differed from last year's—we took the more picturesque western road through Erdene Dalay instead of going through Mandal Gobi. Neither the steppe south of Ulan-Bator nor the desert steppe and semidesert around Dalan Dzadgad has any trees, except for some lone desert elms and for small copses of these trees in some of the valleys and ravines. We grew particularly fond of one tall, spreading elm tree which rose in solitary grandeur in a ravine between Mandal Obo and Dalan Dzadgad. We had often taken this route in 1964, and had always stopped here for a rest. The tree gave the only shade in the vicinity, so this time we again stopped at it for a rest and our midday meal.

In Dalan Dzadgad we rented storage space for our equipment and collections, and left some of the gasoline drums in it right away. Dovchin was also busy, hiring laborers to take to the Nemegt Valley.

By June 5th the last purchase had been made and the unwanted baggage left behind, and we were free to proceed on our way. Since not all the laborers had yet been hired, Dovchin remained behind in Dalan Dzadgad with the Mongolian truck; he was to join us in Altan Ula in two days' time.

We said goodbye to the five-man Bayn Dzak group: Halszka Osmólska (in charge), Jerzy Malecki, Jerzy Lefeld, Józef Kaźmierczak, and Wojciech Siciński. It was agreed that I would come to Bayn Dzak in about three weeks with one of our trucks returning to Ulan-Bator.

The Mongolian truck with Dashzeveg's group had gone straight from Ulan-Bator to Bayn Dzak, where it was to meet the Polish group.

We set out at last, heading west for the Nemegt Valley. The two Star 66s were accompanied by the Tatra truck, loaded to capacity, which was to take our equipment as far as Gurvan Tes in the Nemegt Valley (the Tatra would not have been able to negotiate the sayrs and shifting sand dunes beyond that point).

10. The Altan Ula Camp

... We were particularly fond of the baby camels with their disproportionately long legs ...

The summer of 1965 was unusually cool and rainy, not just in Poland but in the Gobi as well. This was evident to us as soon as we reached Dalan Dzadgad, where the public garden in the center of town looked greener and more luxuriant than before. Once past the Gurvan Saikhan pass and beyond Bayn Dalay, on entering the Nemegt Valley, all those on their second or third trips could notice that the semidesert extending through the center of the valley, which they remembered as being yellowish-gray, was now greenish. We also found other changes. The year before, the township of Gurvan Tes, which lies near a small brackish lake being processed for salt, had been the somon seat. In 1965 the aymak authorities decided that the township's location among the dunes, coupled with the very strong prevailing winds, made it unsuitable as somon seat. The scanty vegetation in the place was not enough to maintain large flocks of sheep and herds of camel. The county seat was therefore transferred to another point in the Nemegt Valley, some 25 miles away to the southwest of Gurvan Tes. Since the somon was still called Gurvan Tes, there were now two localities of that name: the county seat and the salt pan. At the salt pan Gurvan Tes we were greeted as old friends. The local laborers helped us unload the Tatra, and we left a supply of plaster and boards and a number of crates of foodstuffs behind, to be collected a few days later. We left the driver of the Tatra and our kind Gurvan Tes hosts on very friendly terms and resumed our journey.

In 1963 there had been no yurts anywhere to the west of Gurvan Tes. In 1964 we had encountered two or three lone yurts between Gurvan Tes and Naran Bulak. This year, owing to the change in the climate and the unexpected appearance of vegetation, the western Nemegt Valley was well populated by nomad camelherd families and their camels.

Past Gurvan Tes we met a very large camel caravan on its way to Dalan Dzadgad with a consignment of wool. The animals, loaded with heavy sacks, were shorn, displaying a bare grayish skin. In the Nemegt Valley we often met small groups of camels that had escaped from their owners and were living wild in the desert. These camels, with nobody to shear them, were shedding large slivers of fleece which the wind then carried off, and these, mixed with sand, could be picked up all over the Valley. Toward the end of August the camels changed their appearance in preparation for winter, and their still-short brown fleece came to resemble fine, downy velvet.

According to some Mongolian zoologists, there are still wild camels in the Trans-Altaian Gobi; most zoologists, however, now lean toward the view that these "wild" camels are simply domesticated ones that have been living wild for a long time.

We were particularly fond of the baby camels, with their disproportionately long legs. On our visits to the yurts in the Nemegt

Valley, we repeatedly watched the feeding of this younger generation. The baby camels would be tethered; this was enough to ensure that the adult camels would not budge from the place. The owner's wife would lead out one of the baby camels, whose mother would then leave the adult herd and walk toward it. As soon as the baby began to nurse, the woman would attempt to milk the other teat of the mother, collecting perhaps a large cupful of milk from each mother in this way; after the baby camel had had its fill, nothing more could be milked from the mother. Camel milk contains much butter fat and tastes like a better grade of cream. In this nearly horseless part of the Gobi, kumiss is made from camel's milk. We preferred it to kumiss made from mare's milk, as it was creamier and had a finer flavor.

On our way through a sayr in a previously uninhabited area not far from the Naran Bulak area, we passed a well. The well was in a good state of repair, and the water table was very high—just over 5 feet down. A hundred yards from the well stood a lone yurt, inhabited by a family with large herds of camels. As soon as they spotted automobiles from afar, the hostess began to prepare tea.

We became very friendly with the head of the family and with his wife, whose name was Od, and who always aroused our admiration with her engaging manner and hospitality. We would often have tea or kumiss in their yurt when coming from Altan Ula to take on a fresh supply of water at the well, and never failed to admire the cleanliness, tidiness, and good taste which reigned inside the yurt. We would sit on the floor on thick, motley-colored camel-hair carpets. We did not know at that time that our gentle hostess was a deputy to the Great Khural of Mongolia.

Almost 25 miles of tortuous track separated the yurt at the well from our destination, Western Altan Ula. In this area no more yurts were seen. We drove at first along one of the sayrs paralleling the Valley's east-west axis, after which we had to turn north and, willy nilly, cross the dune range extending over several miles. This last, hardest part of the journey was a severe trial for our drivers and took several hours. Finally, on the evening of June 7, eight days out of Ulan-Bator, we reached our destination. We drove up to the cliffs of Altan Ula IV from the west. East of the black, gravel-covered plateau on which we were standing there stretched some ten square miles of Cretaceous sandstone hills, cliffs, and sayrs. Still farther to the east we could see a lower-lying plateau and a new group of sayrs—Altan Ula III—which we had reached from the east the previous year on a short reconnaissance trip from Tsagan Khushu. Our next job was to find a suitable route from the plateau down to the sayrs, since without one we would be unable to bring out specimens; it was easy enough to get down, but not so easy to get back up. Our first try at getting down into the sandy sayr

99 The Altan Ula Camp

Driving in one of the Altan Ula sayrs. The tracks in the sand were made by the trucks of the expedition, no one having ever driven this way before

Altan Ula camp, pitched in the field below the outlet of the sayrs. Cretaceous sandstone hills and ravines can be seen just behind the camp; Altan Ula range is in the background

was a failure: the truck went down all right, but got stuck in the sand and couldn't come up again. Fortunately, all Star 66's are equipped with winches, so the other Star was able to pull the stuck one out. As it was now late, we set up our first camp on the plateau over the sayrs, deciding to try again only after a better reconnaissance of the terrain. Next day we found a sayr with a harder surface and a gentler slope, but even there we had to make a track, by removing the large rocks blocking the way and leveling the sections that were too steep, before we could send down our first vehicle.

In camp the first man up in the morning was the cook who made our breakfast. Reveille was at 6:30 A.M., breakfast at 7, and we would leave for work at 7:30. We would work till 1 P.M. and then come back for lunch. In June the midday break would last for four hours, since not even the toughest of us was able to work between 1 and 5 P.M. In July and August, when the weather became unexpectedly cool—on the whole no more than 90° in the shade—we cut the midday break down to three and then to two hours. In June, however, the "cool" summer notwithstanding, the noon temperature was 100—105° in the shade, the black gravel on the plateau would heat up to 160°, and it was only the almost constantly blowing wind that enabled us to withstand the heat. During the noon break we would rest in the shade. In the desert, "shade" is easily said but less easily found. There was no question of staying in the stiflingly hot tents during the day; our only shade came from awnings put up over the tables and from a narrow space along the north sides of the trucks. We would spread out our camp beds and mattresses there and try to get some rest. Afterward we would go out into the field again and work till sundown, i.e. till 8 P.M. Supper was at 8:30 and lights out—not very strictly observed—at 10 P.M.

Two days after our arrival, Dovchin and two laborers came with the Mongolian group's truck, accompanied by the small Gaz 69 field car hired (with its Mongolian driver) in Ulan-Bator for three weeks. The field car was intended for brief reconnaissance trips in the Nemegt Valley. Dovchin had been unable to hire all the laborers we needed in Dalan Dzadgad, and decided to make a trip to Gurvan Tes (somon) to hire some more. This trip was crowned with success, and we had a group of six Mongolian laborers, four men and two girls, who stayed with us throughout our stay in the Nemegt. We grew particularly fond of Gunzhid and Zorikht, the two girls, who were diligent, strong, and nimble-handed. They soon learned to distinguish between bones and rocks and often went out alone to hunt for bones, with good results. They were very good at gluing the bones together with polystyrene, and learned to number and crate them. The men were also hard-working and strong, and were practically indispensable in moving the heavy monoliths. The

Altan Ula camp, 1965, during a sandstorm. In the picture: E. Rachtan and T. Maryańska

Mongols' camp was about 300 yards away from ours, and housed Dovchin, the laborers, a laboratory assistant named Namsray, and a driver by the name of Dzhamba. Another Mongolian, the geologist Barsbold, arrived at the beginning of July and immediately endeared himself to us through his engaging manner and keen sense of humor. There were thus altogether twenty of us in the Nemegt Valley.

The collaboration with our Mongolian colleagues went without a hitch. Even though some of our members (mistakenly) thought they could speak Mongolian, it would have been very hard to communicate with the local authorities and with our own laborers without Dovchin. I always went over the next day's work program with Dovchin, or in his absence with Barsbold.

A few days after we reached Altan Ula our camp was visited by a sandstorm. Strong westerly winds had been blowing ever since our arrival, giving all of us red-rimmed eyes and the more susceptible of us conjunctivitis. That day, after lunch, we noticed something in the east that looked like a leaden-gray rain cloud but which extended from the ground to the sky. We knew that this was no rain cloud, but rather a sand cloud, presaging a sandstorm. The wind was getting wilder by the minute. I sat zipped up in my tent, with dust and sand blowing in from all directions, and waited for what was coming. Suddenly it became dark, as quickly as in a solar eclipse; a gust of wind tore out the tent pegs and lifted the floor of the tent, turning the heavy bookcase standing on it (which served as our library) upside down; both tent poles crashed down, one of them splitting apart. There was no fear of the tent being carried away by the wind, since it was one of our few tents that had a floor, and this was weighted down by the heavy bookcase. I crawled to the zipper on my hands and knees and managed to get outside. At first I could see nothing but swirling sand; after a while, however, by dint of squinting, I made out a few crate covers whirling in the air above us, with a huge sheet of corrugated packing cardboard sailing high above them, like a kite. All the other tents but one were down; I ran up to one that was about to blow away and lay down on it, together with a few other people, in an effort to save it.

None of us could tell how long it all lasted—perhaps an hour, perhaps longer. As soon as the wind died down we got started restoring the camp, which now looked like a battlefield. Two of us went down to the Mongolian camp to see how it had stood up under the storm; they, too, had no casualties. We spent the next day mending torn canvas, looking for objects carried off by the wind, and generally putting things in order.

Water was a serious problem. The nearest well was at Od's yurt, a mere 25 miles away. Once every five days we had to send a truck on the difficult journey across the dunes to bring back three

50-gallon drums of water. Since we also used the water for plastering, consumption was high.

On Sunday, which was usually not a working day, some of us would go down with the water truck and do our laundry at the well. Others would use the day to make an excursion into the nearby mountains. The Altan Ula Cretaceous deposits were very close to a Paleozoic massif whose foot lay only 4 miles from camp, so an excursion to the highest peak took no more than a day. At Nemegt, on the other hand, where the Cretaceous deposits lay 10 miles from the foot of the Paleozoic range, a trip to the mountains took two days.

The day after he arrived, Dovchin, who was familiar with the Gobi fauna, drew our attention to hoofprints in the sand not far from camp. The prints were not cleft but were smaller than horse hooves, and must have been made by a kulan (*Equus hemionus*), sometimes called a wild ass. The Soviet geographer Murzayev, an expert on the flora and fauna of the Gobi, had reported that he chanced on a herd of kulans a thousand strong near Bon Tsagan lake in 1941. We were very anxious to see a kulan, since they were very rare in our area, but the one in question seems to have sensed our presence and fled. On my two visits to the Gobi I saw a kulan only once, from 300 yards away, as I was driving in the southwest of the Nemegt Valley.

Our constant companion in the Nemegt Valley camps was the large-eared hedgehog (*Hemiechinus auritus*). It was our practice to dig a garbage pit in each camp, some distance away from the tents, and these pits were always visited at night by the hedgehogs. Dobiesław Walknowski, a great lover of hedgehogs, caught one on our second evening in Altan Ula and brought it back to camp, insisting that if we took good care of it, it would become tame and not try to run away. We accordingly put the hedgehog in a box, and Marek Łepkowski—who stayed behind in camp in the morning to prepare lunch—was expected to take it for a walk every day. During meals the hedgehog was let loose; it had in fact become domesticated (or so we thought), and would walk around the table and eat everything we tossed it. At the first opportunity, however—during the sandstorm, which upset its box—it escaped, never to return. We then tried taming two other hedgehogs, but these were shyer, refused to become domesticated, and also escaped eventually.

The camp was always full of lizards, whom our presence did not disturb in the least, and on a few occasions we were visited by vipers. Since we would often walk about in sandals, the presence of a viper in camp was dangerous, and the more cautious members advocated killing any viper found. However, Wojtek Sarżyński, who was our greatest animal lover, would never agree to the wanton killing of

105 The Altan Ula Camp

Exposing the pelvis of a large quadrupedal dinosaur at Altan Ula. In the picture: T. Maryańska (left) and Z. Kielan-Jaworowska

a viper. He let Henryk Kubiak catch the first one and preserve it in alcohol for his animal collection. But when another viper appeared in camp, Wojtek put it in a bucket and carried it over a mile away from camp, thus saving its life.

Prior to our departure for the Gobi we had been warned of scorpions. The Gobi scorpion (*Buthus martensi*) didn't look menacing, however, and we had no case of scorpion bite. Gobi scorpions are grayish-green and small, about an inch long. They usually hide under rocks and flee as soon as the rock is lifted. They found a new favorite hiding place, however, under the floors of our tents. Some of the tents had no floors, but others, my own small one for example, had rubberized flooring. Every time I folded up the tent to move it to a new camp site I would find under the floor a few scorpions, which we would then catch in a matchbox. When our equipment got back to Warsaw in the fall, one of the cases was found to contain a live scorpion. It proved to be a female and soon proceeded to multiply, with the result that several generations of its descendants are now living at the University of Warsaw's Zoological Institute.

Even though we knew that scorpion bite can be dangerous, we bore the animals no ill will. Our hatred was reserved for the solpugs—large spiders of genus *Galeodes*—whose bite, according to Mongolian tradition, is very dangerous. Unlike scorpions, solpugs would come into the tents, so we invariably ran a flashlight over tent walls and sleeping bags before going to sleep.

In our first few days of wandering in the sayrs of Altan Ula we often came across major accumulations of bones. Dovchin found a large fragment of a sauroloph's spinal column the day after he arrived, and this was carefully stored away. There were some surface outcroppings of bones, but these skeletons could no longer be extracted, since the bones had been weathered by prolonged exposure and would disintegrate into splinters or powder when touched. We frequently saw spots that had obviously contained a complete skeleton; if the teeth were preserved we could even identify the dinosaur, but the skeleton itself was most often completely weathered and not removable.

Just as frustrating were the skeletons showing through the rock wall and overlaid by 30 or more feet of sandstone. Even if the skeleton could reasonably be expected to be complete and well preserved, its extraction was technically impossible. We liked it best when a skeleton lay below the surface and was thus protected from weathering to a certain extent, but not too deep—say 3 to 6 feet down. Obviously, deeply buried or not, there had to be part of a bone showing, since we never started digging unless we knew for certain that the bones were there to be found.

Our first major piece of work at Altan Ula was to extract the enormous pelvic girdle of a quadrupedal dinosaur. The monolith containing the pelvis was the biggest we ever cut; it weighed two and a half tons. This specimen, which was found by Maciej Kuczyński, lay under a pile of rock near camp, and it was difficult to tell at the beginning to which part of the skeleton the bones belonged. We next uncovered the sacral bone and two huge ilia, $2\frac{1}{2}$ feet apart. An enormous vertebra 10 inches in diameter could be seen behind the pelvic bones. The other parts could not be uncovered, for fear of damaging the specimen. We had to take it away in one piece, as a single giant monolith. The wooden frame which was put over the pelvic girdle in the rock was 4 feet square. These ilia had a structure quite different from those of known dinosaurs. Their lateral spread indicated beyond doubt that their owner was a quadruped, but beyond that nothing could be said. The crate with the heavy monolith had to be transported downhill, which we managed easily by hand trailer, and then up a steep hill to the trucks. This was more difficult, and took a truck trailer. Wooden boards were laid under the monolith and the crate was slowly pulled up. On top of the hill we built a ramp and had a truck back up under it so that the truck floor was level with the bottom of the crate. The crate was then pushed onto the truck by means of a system of levers actuated by another truck. The monolith was then shipped off to Ulan-Bator with the first consignment of collections.

11. Our Biggest Skeleton

Exposing the skeleton of a large sauropod at Altan Ula, 1965. Ribs visible in the center; femur visible on the left

On our fifth day in Altan Ula, Ryszard Gradziński, who was keeping busy by walking about in the sayrs surveying the terrain for a topographic sketch, returned at midday to announce that he had found large and well-preserved bones in the northern cliffs about an hour's walk from camp. After dinner we set out to inspect Ryszard's bones. He himself couldn't join us, as he had to make a few more measurements near the camp to finish his sketch and so to get on with his own work at last. Kuczyński, who was helping Ryszard with the sketch and knew where the bones were, took us to the spot. At 5 P.M., with the sun still unbearably hot, four of us, Teresa Maryańska, Henryk Kubiak, Maciej Kuczyński, and myself, walked out of camp, reaching our destination more than an hour later. The bones looked very interesting. Lying in a patch of shifting sand on the surface was a long, dark, thick bone over 3 feet long. There were also a number of bone fragments nearby, indicating the presence of another bone now completely weathered. About 10 feet away was another large flat bone, with rounded edges, perhaps a part of the scapula. As soon as the covering sand had been removed (by preliminary sweeping and cleaning with a spatula) we saw that there were more bones a little farther down, and that these were whiter and in a better state of preservation than those on the surface. The arrangement of the bones indicated that the skeleton had been preserved almost intact.

We inspected the topography of the find. Approach by truck was quite out of the question. We were on a very high, rocky plateau from which a number of very steep ravines descended in all directions. We would not enjoy carrying the large reptile several hundred yards down the steep ravines, but it was clear even then that we would get it out, no matter what the difficulties.

We walked back to the camp by another route, looking for the nearest approach by truck. We went down one of the sheer-sloped sayrs and came out onto a wide field, through which we had to walk for another two miles to reach the camp. By that time it was already 9 P.M.; Marek looked like thunder, since we were again late for supper, but the news of the large skeleton soon pacified him. Despite the late hour, I walked over to the Mongolian camp after supper to talk with Dovchin. I told him about the skeleton, and we decided to take a large group there the next day, including a number of laborers, and start working. To cut the time needed to cover the distance, we decided to use the Polish truck and the Mongolian truck in turn to drive the group northward over the flat ground for about two miles. We would then continue on foot; the shortest way was across several ravines, with much clambering up and down. It was none too convenient; the walk took 25 minutes in the morning and almost twice as long at noon, walking back to lunch in the heat of the day. For the first few days we had to make four trips between the skeleton and the camp every day, wasting more than two hours in the process. As we lost much time and energy this way, it would have been desirable to move the camp close to the skeleton, but this was not feasible, since work was proceeding at the same time on the southern side of the cliffs, near the camp site. We had just begun removal of the pelvic girdle of the quadrupedal reptile mentioned in the previous chapter, and the preparation of the monolith required a constant supply of plaster, water, boards, and nails. If we were to move nearer to Ryszard's skeleton, the monolith group would have to be driven to and from work every day, and little would have been gained. We therefore resigned ourselves to the long daily trips between the camp and the skeleton site.

Systematic work on extracting the skeleton began on June 14th. A large supply of spades, picks, chisels, hammers, brushes, and spatulas was brought to the skeleton, along with jugs of freshly made polystyrene glue for gluing and conserving the bones; the glue had to be brought from camp every day, and we used over a gallon per day at first. Those working on the skeleton during the first ten days were usually Teresa Maryańska, Henryk Kubiak, Dovchin and myself, and four laborers. During that time Wojciech Skarżyński, Aleksander Nowiński, Maciej Kuczyński, and two laborers were busy near camp preparing the monolith containing the pelvic girdle.

This was probably the hottest period of the 1965 expedition. The daily temperature varied between 100° and 105° in the shade. The plateau on which we had found our skeleton was uneven. The bones were lying in a pit bounded on two sides by embankments of loosely cemented sandstone and on the third side by a tall mound of sand. We kept exposing bones and working our way down. As a result we soon had to work in a system of trenches—excellent protection from all gusts of wind but no protection at all from the almost vertical shafts of sunlight beating down on us from the cloudless June sky.

The senior Mongolian laborer, Lam Dzal, always carried a large red parasol bleached by the sun. During the worst of the heat we would stop every hour or so and take turns sitting under the parasol. This was the only shade available anywhere near the skeleton.

However, let us begin at the beginning. The work was begun by digging a ditch that was supposed to circumscribe the entire area occupied by the skeleton, as estimated from the size of what was originally showing on the surface. It was highly desirable that the entire skeleton be in fact included in the area surrounded by the ditch, for this would make it possible for us to work our way cautiously in from the ditch without damaging the bones. Our first estimate of the skeletons's size led us to stake out a rectangle of about 25 yards square, which we thought would include the entire skeleton.

While the laborers were digging the ditch, we proceeded to clean and glue the bones protruding from the ground. We soon saw that they were surrounded by shifting sand only at the surface; lower down they were imbedded in sandstone or conglomerate which was occasionally very hard to remove. This meant that extraction of the skeleton would not be as easy as we had thought. The bones that had lain at the surface for a long time were free of the encasing hard sandstone layer, since the latter had weathered and disintegrated; but, unfortunately, the bones had weathered too and much polystyrene and patience were needed to glue the splinters together to preserve them from further damage.

Digging the ditch was fast and easy at first. On the second day, however, we had a small incident. One of the Mongolian laborers accidentally drove his pick into the hand of the 16-year-old girl Zorikht who was working at his side. Zorikht was very brave about it and refused to stop working, but on inspection we found the wound to be deep. We sent her back to camp to see our surgeon. Marek Łepkowski gave her a special welcome, since she was his first real patient, and put in a few beautiful stitches. Zorikht soon recovered and went back to work.

On the same day, as we were digging the ditch, Gunzhid's spade struck a clod of sandstone. The 19-year-old Gunzhid was very friendly with Teresa Maryańska, who was in Mongolia for the second time, knew many Mongolian words, and could in a pinch make herself understood in Mongolian. Gunzhid was a born teacher and gave Teresa lessons in the language while working on the skeleton. As her spade struck the sandstone, she called out in Mongolian: "Teresa, yas bayna" (Teresa, there are bones here). She was right. We cautiously chipped off a piece of the clod, and saw that it contained segments of relatively small bones, probably fragments of the tail vertebrae. This meant the skeleton was much larger than we originally thought. The bones lying to the west were in a sand mound; those lying north, as we were soon to find out, were encased in a 7-foot wall of weakly cemented sandstone. We knew we would have to remove many cubic yards of sand and dig up the sandstone mound fringing the already protruding bones before we had the entire skeleton exposed. The extraction of the bones from strongly cemented sandstone would be more difficult than from sand; we refused even to think of how we would carry the hard sandstone blocks down the plateau to the trucks.

The reptile's identity preoccupied us—meaning Teresa, Henryk, and myself—from the very first. Identifying the group to which a given dinosaur should be assigned is not difficult even for an

Wojciech Skarżyński carrying a 175-lb. fragment of sauropod skeleton to the workshop

inexperienced paleontologist, provided he has at his disposal skull fragments including teeth, or the pelvic or shoulder girdles, or else well-preserved long bones or limb extremities. If, however, the bones are deeply hidden in the rock and there are only fragments showing on the surface, it is easy to make a mistake.

We knew from the start that the skeleton we were exposing was huge, bigger than either the tarbosaur or sauroloph skeletons hitherto encountered in the Gobi. The first bone exposed was the tibia, one of the shank bones; underneath and parallel to it was the other shank bone, the fibula. We took a good look at these enormous bones. They had the structure typical of the large herbivorous dinosaurs called sauropods, but we knew that sauropod development took place in the Jurassic and that such reptiles were very rare in the Cretaceous and almost unknown from the Upper Cretaceous of the Gobi. The Soviet paleontologists had merely reported finding individual fragmentary sauropod bones, and we had not expected to discover a complete skeleton from this group in the Nemegt Valley. However, the next few days confirmed our tentative assumption. By the time we had uncovered an enormous femur, a part of the highly damaged pelvis, a few spinal vertebrae with very long (over 5 feet) ribs, and a scapula, we could be certain that this was indeed the skeleton of a sauropod.

This sensational discovery was compounded by the fact that a very well-preserved sauropod skull was found at almost the same time during a short trip to Nemegt. On June 15th Teresa, Ryszard, Maciej, and Dovchin, accompanied by a driver, took the Gaz 69 field car for a two-day trip to Nemegt. On their return they reported that Maciej had found a skull encased in a wall of hard sandstone in one of the Nemegt sayrs. It was partly hidden in the rock, but its protruding, stump-like teeth and its shape allowed Teresa to identify it as that of a sauropod. Removing the skull was not easy, since more than three feet of dense sandstone had to be picked out to reach it and the party had not brought the necessary tools. One of our trucks was slated to go to Gurvan Tes the next day to bring back the boards and plaster which had been left behind. We decided that Walknowski, who was driving the truck, would take along Kuczyński—the discoverer of the skull—and Skarżyński and that they would visit Nemegt and extract the skull on the way. They armed themselves with picks and crowbars and three days later returned with the boards, and with the rare, beautiful skull carefully wrapped in wood shavings and cradled in a spacious box.

My three weeks' stay in Altan Ula was drawing to a close, and the time was approaching for me to visit the Bayn Dzak group; we had been out of contact with them for three weeks and were curious to know what progress they had made.

Meanwhile, extraction of the sauropod skeleton was far from

A sandstone block containing sauropod bones being dragged to the workshop

The ladies at the Altan Ula camp. Left to right: Z. Kielan-Jaworowska, Zorikht, T. Maryańska and Gunzhid, 1965

complete. According to plan we were to leave Altan Ula on June 20th together with Kuczyński and Walknowski, to take our first batch of collections to Ulan-Bator. The truck was to drop me off at Bayn Dzak on the way. Kuczyński and Walknowski were to stop over in Ulan-Bator for several days in order to hire another Gaz 63 truck with a Mongolian driver to carry our collections between the Nemegt Valley and Dalan Dzadgad in late July and August. Both trucks were then to collect the rest of the packing material left in Ulan-Bator, call at Bayn Dzak on the way, and return to Altan Ula with the Bayn Dzak team and myself. It was clear I would be away from Altan Ula for at least two weeks, and would thus have to part from my sauropod for quite a while.

Preparing and transporting the pelvic girdle monolith took more time than we expected, and we were only able to leave Altan Ula on the afternoon of June 23rd. By that time a large part of the sauropod skeleton lay exposed. Ryszard made a detailed 1:20 drawing of the exposed bones showing each one's serial number. The serial numbers were marked several times on each bone with a marking pencil. If the bone was disintegrating or threatening to disintegrate, we drew perpendicular lines in China ink or marking pencil across the cleavage. This is very useful in reconstituting a broken bone brought back from the field.

A few days before my departure we began removing the numbered bones. The long bones, which had lain at a shallow depth, were on the whole fairly well preserved. Even so, none was removed whole; they usually broke laterally into several segments, which we thought would be easy to reconstruct in the laboratory. This was also true of the ribs. It was more difficult when we came to the vertebral column and to the pelvic and pectoral girdles, which were imbedded in large blocks of strongly cemented sandstone and could not possibly be extracted in situ. We therefore decided to remove them together with the enclosing rock; this had the advantage of protecting the bones during transport.

We also set up a carpenter shop and a packing station on the nearest spot accessible by truck. The workers would carry the numbered fragments to the shop, where crates would be made to measure and the bones carefully packed in lignin, shavings, corrugated cardboard, or plaster.

I left Altan Ula with a bad conscience, fully appreciating the difficult task facing the team staying behind. It turned out, however, that Ryszard Gradziński, who was my deputy on the expedition and took charge of the Altan Ula camp on my departure, was more than equal to the task of removing and transporting the sauropod. The day after I left he cut the Gordian knot and moved the camp, lock, stock, and barrel, to the vicinity of the skeleton, the new site being a large open field at the issue of the ravines leading to the

Carpenter's shop for crating and packing sauropod bones at Altan Ula camp. Large sandstone blocks with bones are seen in foreground

place where the bones were found.

The distance between the new camp site and the carpenter shop—later referred to as "The Café" on account of the picturesque colored canopy spread over it as a sunshade—was a 5 minutes' walk; the uphill distance from the carpenter shop to the skeleton could be walked in 15 minutes. Aleksander Nowiński was put in charge of the shop. During the first two days at the new campsite more of the arched tail was exposed in the wall (its tip had not been preserved), along with the right hind limb. The skeleton was found to be almost complete; however, the cervical vertebrae, the skull and the last vertebrae of the tail were missing.

Once the skeleton was exposed and quartered, the parts were carried to the carpenter shop. The steep track between the skeleton and the workshop was 640 yards long. The smaller fragments, such as the laterally fractured long bones, were carried by the laborers on their backs. But in addition to these partly prepared bones there were 18 bone-encasing large sandstone blocks to move, the largest of which weighed some 1500 pounds. The problem was how to drag them along the steep path to the workshop. It occurred to Wojciech and Aleksander to cut one of our 50-gallon gasoline drums in half lengthwise and attach stout ropes to make a kind of sled. The block was loaded onto the sled and then drawn along. It took every single member of the Polish and Mongolian teams at Altan Ula to drag the heaviest blocks to the truck. The only trouble was caused by the possessors of cameras, who wanted to immortalize this mode of conveyance for posterity but who couldn't let go of the ropes. The problem was solved by entrusting the picture-taking to Teresa, who had to handle eight ordinary cameras and one movie camera at the same time and could never remember which particular ones she had already used.

Against all expectations, the entire sauropod skeleton had been removed and brought to the workshop by the time I returned to Altan Ula on July 8th together with the Bayn Dzak group. The extraction of the 12-ton skeleton and its removal to the shop had been accomplished in the record time of not quite four weeks.

We couldn't crate the entire skeleton on the spot, since this would have used up practically the entire supply of boards in Altan Ula, and left none for a number of other skeletons we were extracting at the same time. Therefore, after long discussion we decided to crate only the more delicate and fragile parts in situ. These took up twenty crates. The heavy, hard blocks of sandstone with the enclosed bones were trucked to Dalan Dzadgad uncrated. The blocks were hard enough to prevent any damage to the bones inside. We lined the truck floor with saksaul *Haloxylon* branches to act as shock absorbers, and placed the blocks on top of the branches and between the crates. During July our trucks made the 250-mile trip between

121 Our Biggest Skeleton

Altan Ula and Dalan Dzadgad six times in transporting the sauropod skeleton to our collection depot. The blocks were later crated in cases brought in Dalan Dzadgad and then safely shipped via Ulan-Bator to Warsaw.

12. The Cretaceous Mammals of Bayn Dzak

Upper Cretaceous sandstone outcrops at
Bayn Dzak. Flaming Cliffs

Cretaceous Mammals of Bayn Dzak

I first saw the Cretaceous sandstone outcrops at Bayn Dzak in 1964, on my way back from the Nemegt Valley. We had been passing Bayn Dzak at sunset, and I could see for myself how right the American paleontologists had been in calling them the Flaming Cliffs: in the red light of the setting sun they really seemed to be on fire.

Undoubtedly, the American Gobi expeditions' greatest discoveries were the primitive insectivore and multituberculate mammal skulls which they found in the Cretaceous deposits of the Djadokhta formation at the foot of Bayn Dzak's Flaming Cliffs (in the American literature Bayn Dzak is called Shabarakh Usu, after a small lake in the vicinity).

In 1922 the American paleontologists found a multituberculate skull at Bayn Dzak, a skull which Simpson described in 1925 under the name *Djadochtatherium matthewi,* after the Djadokhta formation in which it was found. The next expedition, which took place in 1925, renewed the search for mammal fossils here with redoubled vigor, the result being that eight specimens, including seven skulls, were collected; these were described in 1926 by Gregory and Simpson, who placed them in four new genera of insectivores—*Zalambdalestes, Deltatheridium, Hyotheridium,* and *Deltatheroides.* Of these only the first two were represented by skulls in a fairly satisfactory state of preservation, the other being recovered only in fragments.

The Bayn Dzak beds are thought to be late Cretaceous. However, the late Cretaceous period extends over some 35 million years. While the Nemegt Valley Upper Cretaceous deposits belong to the period's youngest strata (Campanian or Maastrichtian), the Upper Cretaceous deposits at Bayn Dzak are much older, dating back to the middle of the upper half of the Cretaceous (Coniacian-Stantonian), i.e., to an estimated 90 or 95 million years ago. The Bayn Dzak Upper Cretaceous deposits contain not only mammals but also numerous skeletons of small land dinosaurs, mostly *Protoceratops* (the ancestor of the horned dinosaurs), armored dinosaurs, and small carnivorous dinosaurs. These formations also contain large numbers of reptile eggs.

Because of the political changes that took place in Mongolia after 1925, American expeditions were no longer able to enter what was now the Mongolian People's Republic and had to limit themselves to Inner Mongolia, today a part of China. The number of Cretaceous mammals found at Bayn Dzak therefore did not increase.

The Cretaceous age of the Bayn Dzak mammals has been questioned by the Soviet paleontologist Novozhilov. The 1948 Soviet expedition had spent 11 days at Bayn Dzak while Novozhilov made geological observations in the field, but failed to find any mammals there. However, in view of the published American reports that the skulls were from concretions found at the foot of the Flaming Cliffs,

Novozhilov concluded—by comparing Bayn Dzak's geology with that of Khashaat (Gashato in the American literature), which lies 5 miles away and displays earliest Tertiary (Paleocene) deposits—that the concretions in which the Bayn Dzak skulls had been found originated by secondary deposition from Paleocene beds once present at Bayn Dzak but now entirely eroded away.

Novozhilov's results, published in 1954, produced consternation. Admittedly, since he himself had failed to find mammal fragments in the area, his observations could not be considered fully conclusive, but the geological description of the Bayn Dzak profile given by the Americans was none too accurate either, and without knowing the outcrops it was impossible to decide one way or the other. The question remained open for 10 years, and all the paleontology textbooks that appeared during that time necessarily had to place a question mark after the Cretaceous origin of the Bayn Dzak mammals.

It is therefore obvious that the problem of the Bayn Dzak mammals was of great interest to our expeditions.

We had obtained important results during our 1964 stay at Bayn Dzak. In five weeks of crawling on all fours for 12 hours a day over a large field of concretions at the foot of the Flaming Cliffs, a party of our paleontologists had managed to collect 9 specimens of mammals. Dr. Lefeld, one of the geologists at Bayn Dzak throughout this period, made very exact stratigraphic observations of the area and concluded that the skulls were undoubtedly Cretaceous. Over a fairly large area at the foot of the Flaming Cliffs there are outcroppings of horizontal beds of red sandstone. As they weather, these beds leave behind on the surface various-sized lumps of more strongly cemented rocks—lumps that contain bone fragments. It was in such loose lumps that the American paleontologists had found the skulls of their very primitive mammals, and our own finds in 1964 came from the same source. Most of the mammal specimens we found in 1965 were also from these concretions, but we also found a few specimens in situ, i.e., in the standstone bed itself, alongside with dinosaur eggs and *Protoceratops* teeth. This established the Cretaceous age of the Bayn Dzak mammals beyond any doubt.

Prior to our departure in spring 1965, the Warsaw Institute of Paleozoology of the Polish Academy of Sciences was visited by Professor Malcolm McKenna of the Museum of Natural History in New York, an expert on Mesozoic and Early Tertiary mammals. Professor McKenna had come to Warsaw to study our collection of Cretaceous mammals. He offered to take one of our most delicate and difficult-to-prepare skulls to New York to be prepared by one of his Museum's most experienced technicians, and I gladly accepted (the skull was returned to us in 1966 fully prepared).

Professor McKenna also showed us a few of the slides the American expedition had taken at Bayn Dzak. They were in color, even though a color process had not yet been invented in the 1920s (they were actually black-and-white diapositives hand-tinted by Chinese artists). When we produced our own color shots from Bayn Dzak we could see that the colors were very much alike in the two series of slides. This comparison was very instructive, since we could now be sure that we were working in the same area.

Soon after Professor McKenna left we received a precious gift from the archives of the American Museum of Natural History in New York—copies of topographic maps of the Bayn Dzak vicinity made by the American expeditions' professional topographers. For us these maps were priceless, since our team did not include a topographer, and the topographic maps drawn by our geologists were more sketches than maps, and far less accurate than the American maps.

On the evening of June 23rd, 1965, we left Altan Ula for Ulan-Bator with our first batch of specimens. There were three of us—Dobiesław Walknowski, Marciej Kuczyński, and myself. Walknowski drove very fast, and we reached Bayn Dzak late at night on June 24th, after a 28-hour drive broken only by a short night's rest and by a two-hour stopover at Dalan Dzadgad for mail and refueling. The expedition was camped on the open steppe, with the Polish group's tents pitched quite close to the precipitous slopes of the Flaming Cliffs. The field of concretions we were exploring could be reached by simply going down the slope for 20 yards or so. About 300 yards to the west lay the camp of the Mongolian paleontologists. Dashzeveg, Erdenibulgan, a geology student named Galsan, and the driver had established themselves very comfortably in two yurts they had borrowed from a nearby village.

We made an attempt to find traces of the American expedition which had worked the area some 40 years earlier. Dashzeveg scored the first success, learning that a man in the village from which the yurts came had worked as a laborer with the Americans 43 years before.

Wojciech Siciński discovered another clue, a bottle cap printed in English, while inspecting what we calculated should have been the site of the American camp. And that was that, though we did find a number of very rusty food cans left most probably by the Soviet expedition of 27 years later. On my arrival, despite the late hour, I demanded to see the mammals found so far. It turned out that three weeks of laborious search had unearthed six specimens of Cretaceous mammals, including three well-preserved skulls and three lower-jaw fragments. These were quite satisfactory results, considering that our team of the year before had already gone over the site thoroughly and that some of its members had tried to

Incomplete protoceratops skeleton from Bayn Dzak encased in sandstone

Dinosaur eggs in Bayn Dzak

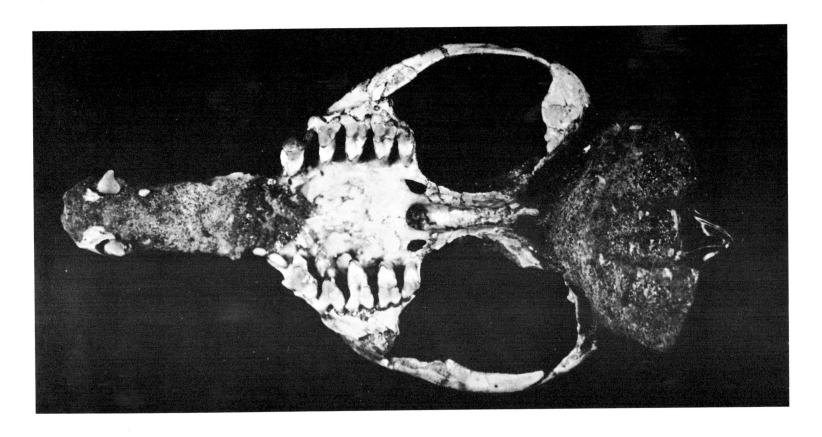

Skull of Cretaceous mammal *Zalambdalestes*, in occlusal view ×3.5

convince us that there was nothing much more to be found at the foot of the Flaming Cliffs.

The next morning Walknowski and Kuczyński left for Ulan-Bator, leaving me with the Bayn Dzak team. Compared to the heat that had reigned at Altan Ula for the past few days, and after the long daily walks from the Altan Ula camp to the sauropod skeleton, the first few days at Bayn Dzak seemed like a vacation on the French Riviera. It was somewhat cooler than in the Nemegt Valley, and five miles from camp stood a shepherds' settlement with a few real houses and even a store. Ten miles away was the village of Bulgan, whose mountain spring formed a small waterfall that could be used for bathing. Bulgan had a post office, a bakery, a number of stores, and an experimental breeding farm belonging to the Agricultural Station of the Mongolian Academy of Sciences. We were thus quite close to civilization.

The Bayn Dzak finds made prior to my arrival included, in addition to the mammal material, several small lizard skulls from the same strata and two *Protoceratops* skulls. There were also two incomplete *Protoceratops* skeletons, a beautiful nest of dinosaur eggs, and skeletal fragments from various reptiles.

On my second day in Bayn Dzak we had an unexpected visitor, Professor Tserev, Vice-President of the Mongolian Academy of Sciences, a histochemist who was visiting the Bulgan experimental station and had dropped in at our camp on his way. We showed his party the unimpressive specimens we had found, expressing our regret that we could not take them to Altan Ula to see the gigantic sauropod and the tarbosaur skeletons.

Whereas at Altan Ula the main working difficulty was the skeletons' enormous size, at Bayn Dzak it was their smallness. The only way to find mammals was to crawl on all fours with your face close to the ground and to inspect each piece of sandstone under a magnifying glass. Jerzy Małecki thought of arranging the lumps already inspected in rows so as not to waste time looking at the same specimen twice. Since the field at the foot of the Flaming Cliffs (which we named the Main Field) seemed exhausted, it had been decided a few days before my arrival to look for other "mammaliferous" grounds. We therefore moved to a point a mile and a half to the west, where the same horizon outcropped over a small area. This new field was called the "Ruins," since it consisted of a group of red sandstone strata which from a distance resembled an old castle. Though not large, the field had already yielded the lower jaw of an insectivore, and so could be expected to produce more. We therefore began a systematic exploration of the Ruins, and got very satisfactory results. Halszka Osmólska, a mammal collector of enormous patience and concentration, found a beautiful and quite complete multituberculate skull of a structure quite unlike

Skull of Cretaceous mammal *Kennalestes* in lateral view, ×7

131 Cretaceous Mammals of Bayn Dzak

that of other representatives of the group and certainly belonging to a new family. We also found three lizard skulls, along with a few incomplete mammal skulls (mostly lower jaws).

Our third hunting ground at Bayn Dzak lay a mile and a quarter to the west of the Ruins. We named it the "Volcano," after a volcano-shaped mountain nearby. Here, as at the Ruins, the mammal-bearing sandstone outcropped over a very small area only, but, since nobody had been there before, a few satisfactory specimens were found, including three well-preserved *Zalambdalestes* skulls. This was very gratifying, since all the mammals we had found so far were forms different from those the Americans had found in the area, so different in fact that they represented entirely new genera. This was an indication of considerable mammalian specialization during the Cretaceous, but we had nevertheless been upset at not finding any specimens of the genera represented in the New York collections. The *Zalambdalestes* specimens from the Volcano could now be considered the (no longer) missing link between the two collections. One of the new Cretaceous eutherian mammals from Bayn Dzak I described in 1968 as *Kennalestes gobiensis*.

By the time we broke camp at Bayn Dzak on July 6th we had put together a collection of more than a dozen Cretaceous mammal specimens. We could not tell the exact number, since some of the specimens were encased in rock, and we would not know whether they were mammals or lizards until after they had been prepared. Back in Warsaw, preparation subsequently showed that we had 21 mammal specimens, including 9 very well preserved skulls. The 1964 and 1965 expeditions thus yielded a total of 30 mammal fossils.

The obvious question at this point was why mammals had been preserved at Bayn Dzak but not in the Nemegt Valley. The answer was quite simple. The Nemegt's Cretaceous deposits were formed under water; they were lake or river deposits. The Nemegt Valley horizons, which contain large accumulations of dinosaur skeletons, were probably deposited by floods which caused mass slaughters of dinosaurs. The Bayn Dzak Cretaceous sandstones, on the other hand, are windborne deposits probably formed under conditions similar to those now prevailing in semi-arid southern Mongolia. Countless numbers of small mammals live there today, including rodents and insectivores, and there are (on the nonmammalian side) small lizards. The zone is also inhabited by large animals adapted to the arid climate, which in our era are of course mammals, mainly camels and gazelles.

A hundred million years ago, when the sandstones of the Djadokhta formation were being deposited, such semi-arid zones were inhabited by primitive insectivorous mammals and by multituberculates, the latter having occupied the niche of our modern rodents.

133 Cretaceous Mammals of Bayn Dzak

Decorative Mongolian tents erected for the 1965 Nadom at Dalan Dzadgad

There were also large animals adapted to the dry climate—at that time quadrupedal land dinosaurs such as *Protoceratops* and the armored dinosaurs.

While at Bayn Dzak we made several short trips to the locality of Khashaat, 5 miles away, where there are Paleocene deposits bearing a mammalian fauna. According to the American paleontologists these formations contain primitive mammals similar to those we had unearthed at Naran Bulak. However, while we were distinctly lucky at Bayn Dzak, we were out of luck at Khashaat. All its Paleocene outcrops had been concealed by recent sliding. We saw a system of small hills, all exactly alike, covering several square miles. In two of these hills the Americans had found lenses containing mammalian fossils, and the subsequent Soviet expeditions had made similar discoveries, but we for our part could find nothing. The only fossils we could bring back from Khashaat were a few unidentifiable fragments.

Thin layers of calcareous concretions are sometimes found in the Gobi's sandstone series, both in Cretaceous and in Tertiary deposits. We collected these very carefully, since we hoped that after dissolution in acid the enclosed plant pollen could be isolated and the age of the layers thus established. Dr. Henryk Makowski, the geologist on our first reconnaissance expedition, brought back the first collection of these concretions. Much to our disappointment, they contained no traces of organic matter.

On July 4th, 1965, the Starr 66 brought Kuczyński and Walknowski back from Ulan-Bator to Bayn Dzak. The rented Gaz 63 truck, with its Mongolian driver Izh and with the new Mongolian member of the expedition, the geologist Barsbold, arrived on the following day. By July 6th camp was struck at Bayn Dzak and we left for Altan Ula. That day Dashzeveg invited us to his yurt for a farewell party; he was not going along to Altan Ula, since Nadom, Mongolia's national holiday, was coming up, and he and his group did not want to miss the great celebration in Dalan Dzadgad. We therefore decided that the Mongolian group from Bayn Dzak would stay about a week in Dalan Dzadgad, where it could join up with the Polish group leaving for Western Mongolia. Though we greatly regretted missing the Nadom festivities, we were forced to return to Altan Ula to see what progress had been made, especially as the final completion date for the field work had been moved up. However, the celebrations of two years before had been witnessed by the members of our 1963 reconnaissance expedition, and Maciej Kuczyński and Andrzej Sulimski had given the following description in a 1964 journal publication:

"The three-day celebration of Mongolia's national holiday began in the capital on July 10th. There was a military parade in the central square and a procession of civilians who had come in from all over the country. At the same time the traditional Mongolian competitions began at different points in the city. The most important sport is horseback riding, consisting of men's and boys' horse races run over tens of miles. Even 9-year-old boys and girls were riding over a 20-mile course, screaming their heads off and belaboring their horses with their whips. On the sports ground a special kind of wrestling was going on. The opponents were in jackboots, shorts, and brightly colored, richly embroidered vests, with seconds shouting loud encouragement. The first one to touch the ground with any part of his body other than his feet is the loser. The winner puts on a conical cap and runs up to the grandstand to do the dance of the victorious eagles. Crowds of Mongolians were also watching the archery competitions. The aim is to hit a square marked on the ground from some 50 yards away. After each hit the umpires all lift their hands over their heads and turn around, singing as they turn."

13. There Ain't No Such Animal

. . . we sat by the fireside . . . listening to
beautiful Mongolian songs . . .

We were on our way to Altan Ula again, taking the familiar route through the desert sands of the Nemegt Valley. Four of the five persons from the Bayn Dzak crew were going to the Nemegt Valley for the first time, and we did our best to give them a fright by telling them of the Spartan conditions at the camp.

Past Gurvan Tes we saw one of our trucks on the skyline. This was the first consignment of crates with the sauropod skeleton being taken to Gurvan Tes by Rachtan and Maryańska.

We got to Altan Ula on July 8th, and the camp immediately became very noisy. We were all together again for the first time in six weeks, i.e., ever since we had split up in Dalan Dzadgad. We took advantage of this to take another picture of the Polish group for posterity; this one turned out to be more interesting than the one taken on the way from Ulan-Bator to Dalan Dzadgad, since most of the gentlemen had grown beards by now.

With the Mongolian national holiday two days away, our Mongolian colleagues invited us to help them celebrate. On the afternoon of July 10th, instead of going out to work, we walked over to the Mongolian camp for the party. Lam Dzal, one of the Mongolian laborers and an excellent shot, had killed an argali (a wild ram), and we were served mutton with rice, alcohol made from milk, kumiss, and other Mongolian delicacies. After supper the traditional Nadom competitions began, including a marksmanship contest, and Ryszard Gradziński astounded us all by walking barefoot on the glowing coals of the fire; it turned out he had learned to do this in Bulgaria. We then sat by the fireside for a long time, listening to beautiful Mongolian songs.

The following morning we woke up late, and found it was raining. We looked in amazement at gray skies and at mountain peaks shrouded in mist. It was raining so hard that we couldn't go out to work all morning.

By this time work on the sauropod skeleton was finished and we were uncovering a few other skeletons, all at the same time. Wojciech Skarżyński had found a big tarbosaur skeleton in the Altan Ula III outcrops a few miles west of camp. Unfortunately, the skeleton lay in a sandstone layer which outcropped on the side of a ravine. A few large sandstone blocks had fallen to the bottom of the ravine and destroyed some of the bones, but the bulk of the skeleton, including the front of its huge, 50-inch-long skull, was in fairly good condition. We drove out to Altan Ula III the next morning to bring the tools and begin work. As we were again caught in a pouring rainstorm, we were unable to continue removal of the skeleton, since this involved continual impregnation of the bones with polystyrene. We therefore decided instead to reconnoiter the area and to re-explore the Altan Ula III outcrops, which we had so far failed to examine thoroughly enough.

Polish members of 1965 expedition in Altan Ula camp. Left to right: M. Kuczyński, A. Nowiński, W. Skarżyński, E. Rachtan, D. Walknowski, H. Kubiak, T. Maryańska, J. Małecki, R. Gradziński, J. Lefeld, H. Osmólska, M. Łepkowski, Z. Kielan-Jaworowska, J. Kaźmierczak, W. Siciński

While I was walking in the rain along the southern part of the outcropping, I suddenly noticed some fairly large bones showing on a small hill. There were more than a dozen of them, and they were lying at a fairly shallow level. One protruding bone was about a yard long and rather thin. Nearby, a few large phalanges almost a foot long were sticking out of the sand. I began removing the sand from around the bones. Suddenly I saw there in the sand an excellently preserved, powerful, strongly arched claw twelve inches long. It was undoubtedly a forelimb claw, but larger than any ever found before. The predatory dinosaurs known to have roamed the Gobi, the tarbosaurs, had very short and shrunken forelimbs whose claws were never longer than two inches, even in the largest specimens. The longer claws on the tarbosaurs' hind legs were never longer than four inches, in other words only a third the length of the claw I had just unearthed. The large herbivorous sauropods also had clawed limbs, but their claws were much smaller and quite differently shaped. The partly bare protruding bone looked like an arm bone; if this was actually the case it meant I had come across a long forelimb bearing claws bigger than any ever seen before.

I looked around for someone to share my sensational discovery. Far away against the horizon I saw the bent silhouette of Wojtek Skarżyński. I called out to him, and we began to remove the sand together. It was time to go back for supper. We left all the bones in their original positions so as to be able to make an accuate sketch of their disposition, and started back to camp. There at supper I tried to tell the story, but my listeners were incredulous. A claw a foot long? There was no such animal!

On the following day we drove back to Altan Ula III with a larger group. A number of persons had been entrusted with systematic work on Wojtek's skeleton, whereas Halszka, Barsbold, Edmund, and myself went to dig out mine. The work progressed very quickly. The bones were imbedded in loose sand but were very well preserved. After a few hours of strenuous work we had exposed a narrow and very long shoulder bone and a coracoid plate. The length of the shoulder bone, which was fused with the coracoid, came to 5 feet. At an angle to it was the (originally protruding) yard-long arm bone, as well as two forearm bones and numerous scattered phalanges. Scattered about on the surface about 10 feet from the exposed shoulder bone were a number of bone fragments, from which we managed to reconstitute the other shoulder bone and coracoid along with parts of a forelimb. By the following day we had managed to expose all the bones of the shoulder girdle and of the two forelimbs. We found three claws for one limb but none belonging to the other. The carpals were also missing, but otherwise the limbs were complete. We also found

Huge forelimbs and pelvic girdle of an unknown dinosaur from Altan Ula III. Foreground, left: scapula fused with coracoid bone; in the center: humerus

a few rib fragments nearby. Further digging unfortunately remained fruitless; the shoulder girdle and forelimbs were all that remained of this strange dinosaur, the remainder having disintegrated long ago. After the original position of the bones had been accurately sketched and the bones numbered and removed, we attempted to calculate the probable length of the forelimb. The result was astounding: these forelimbs, armed with long sharp claws, were 8.5 feet long.

That evening I sat in my tent a long time with Halszka, Teresa and Henryk, going over all the literature we had brought with us. The skeleton could not be assigned to any known family of dinosaurs. The positioning of the fingers in the forelimb was the same as in the saurischian dinosaurs belonging to order Theropoda, which includes the carnivorous dinosaurs and the ostrich-like dinosaurs. Our individual, however, differed from the known representatives of both these groups in many respects, first and foremost being its gigantic size. Undoubtedly we had found a representative of not merely a different species or genus, but of an altogether new family of theropods.

On July 12th we split up for a second time, with some of us, Halszka Osmólska, Maciej Kuczyński, Henryk Kubiak, Józef Kaźmierczak, and Dobiesław Walknowski, leaving for the west; this time we were to be separated until the end of the expedition. Two other trucks left for Dalan Dzadgad at the same time: the Mongolian truck driven by Dzhamba and the rented one driven by Izh, with a further consignment of specimens aboard. The trucks were accompanied by Dovchin.

On the following day the camp seemed empty, and we all felt a little strange. We didn't know how well the "western" group would succeed, whether they would manage to reach Altan Teli or perhaps encounter fresh, unforeseen difficulties. We were not in a position to keep worrying for long, however, since there was still a lot of work to do at Altan Ula itself. Work on Wojtek's skeleton was all but finished, and we had also succeeded in crating the huge forelimbs of the strange dinosaur from Altan Ula III, when Edmund Rachtan, taking a lone walk during one of his trips to Altan Ula III, found another large tarbosaur skeleton. Luckily, the site was accessible to trucks, so there were no major difficulties in the extraction. After our experience with the sauropod skeleton, removing a tarbosaur skeleton "only" 40 feet long was child's play.

But this was not all. While I was at Bayn Dzak, a party had made a two-day reconnaissance of Tsagan Khushu to take another look at the old familiar home of last year's expedition. On their walks through the sayrs Skarżyński and Gradziński had found an accumulation of large bones—the skeleton of yet another tarbosaur. We now decided to send a party of eight to the site for a few days to extract it. The bivouac party to Tsagan Khushu consisted of Teresa, Ryszard, Wojciech, and Aleksander, as well as a Mongolian group made up of Barsbold, Namsray, Gunzhid, and Ongoy. They took enough food for a few days and a supply of plaster and boards for the skeleton. The truck that took them out returned that evening to Altan Ula with a fresh supply of water. We did not know at the time how complete the protruding skeleton would turn out to be or whether enough crating material had been taken to cast all the monoliths and crate all the bones. As a result we decided that after two days' work on the Tsagan Khushu skeleton Ryszard Gradziński would send me a progress report, including a request for additional packing material if necessary. But how were we to comply with this? Altan Ula and Tsagan Khushu were only 10 miles apart, but the route included a dune belt several miles wide, which took as long to cross by truck as it takes to cover the [200-mile] distance between Warsaw and Cracow by car. In any case, the trucks were badly needed for work in Altan Ula, and we could not afford to send one to Tsagan Khushu just to bring a message. Another way had to be found. Now, from the top of the tallest hill in the vicinity of the Altan Ula camp there is a panoramic view of the southern part of the Nemegt Valley. The wide belt of dunes with the tall Cretaceous sandstone buttes of Tsagan Khushu can be plainly seen. The Altan Ula hills themselves are just as readily visible from the heights of Tsagan Khushu. We therefore agreed with Ryszard that in three days' time, when the work in Tsagan Khushu should be in an advanced stage, we would make a signal fire in the evening. At 9 P.M. sharp the Tsagan Khushu group was to light a fire and then extinguish it. We for our part were to answer the signal by lighting a previously prepared pile of *Haloxylon* on the hill. A second fire lit in Tsagan Khushu would mean a request for fresh supplies of plaster, boards and food to be sent the following day. There was no second fire, which meant that work at Tsagan Khushu was in its final stages and that the amount of packing material available was sufficient. On the appointed day, four days later, I drove out to Tsagan Khushu in the Star 66, accompanied by Rachtan. The entire skeleton of the big tarbosaur had been extracted and packed in the truly record time of four days. We took the tarbosaur crates to the yurt near the well and returned to Altan Ula with a new supply of water.

Despite the fact that during the second half of July two trucks kept shuttling between Altan Ula and Dalan Dzadgad with our collections, there were so many crates to be shipped that their number didn't seem to be getting any less and it looked as if we would never get them all out of Altan Ula.

In the last few days I jokingly begged my colleagues not to find any more big skeletons, since we would never be able to take them

Tarbosaur lower jaw extracted by J. Małecki

Moving camp from Altan Ula to Nemegt.
Driver: E. Rachtan

back. Toward the end of our stay at Altan Ula, Siciński did find a small ornithomimid skeleton quite near camp, unfortunately without the skull, and spent a few days extracting it, while Lefeld found the fragmentary remains of a large armored reptile. In a loose rock in one of the sayrs Jerzy Małecki found the pelvis of a large tarbosaur; it took up two enormous crates. Małecki also spotted a bone outcrop in the cliff face of the ravine, some ten feet above the bottom. There was no approach, since the bones were covered by another 30 feet of sheer cliff. But Małecki would not give up; he knocked together a ladder out of a few boards and proceeded to pick out the bones while standing on the ladder. He was rewarded with a big, beautiful tarbosaur lower jaw containing a row of small sharp teeth with secondary serrations. It should by now be obvious to the reader that the find was baptized "Małecki's jaw," after its discoverer.

The July 22nd [Polish] national holiday was celebrated with great ceremony at Altan Ula. We invited our Mongolian colleagues to the party, lit a campfire in the evening, and played the traditional fireside games.

The last consignment of crates was sent off to Altan Ula on August 25th, after which we all gave a sigh of relief. We could now get on with dismantling the camp.

Moving day was July 28th. A few days earlier Dovchin informed us that we were all invited to a party at the yurt near the well. For the last time we rode past the well that had been our water supply for almost two months; in the Nemegt we would have another well, somewhat closer at hand.

Our friends—Mrs. Od and her family—were giving us a farewell party in their yurt. Twenty persons from the Altan Ula camp, Poles as well as Mongolians, were invited to dinner, which was boiled mutton with macaroni, kumiss, dessert, and tea. Mrs. Od took new Mongolian dresses out of the chest and asked us to put them on for a group photograph. In Mongolia, men and women dress alike, in trousers and long leather boots. Over this they wear the dela—a kind of overcoat, worn on the left side, made of colored and often glossy or embroidered silk or cotton material with a linen belt of a different color. The lining of the dela is the same color as the belt, and this color is also hemstitched around the cuffs and the fastening. In winter, woolen delas are worn. The head is always covered.

It would be difficult to describe the cordial atmosphere of this last visit to these dear friends from whom we were to be parted for a long time.

14. Nemegt, 1965

Nemegt camp, 1965

We devoted our first couple of days at Nemegt to setting up camp, including tents and facilities. We dug a spacious cellar in the northern slope of one of the sayrs for our supply of canned food, built a kitchen, dug a garbage pit, and finally knocked together a table and a few benches over which we spread kitchen awnings.

During these two days we also explored the area. Nemegt seemed richer in finds than Altan Ula; single bones, phalanges, vertebrae, and fragments of long bones were found frequently. It was enough to take a few hours' walk to return to camp with a bagful of bones. On our first day in Nemegt, Małecki found the hind limbs and a few tail vertebrae of a small ostrich-like dinosaur near camp. They were lying in the sand close to the surface, so it only took him a few hours to extract them. However, we began systematic exploration on the third day, after our camp had been fully set up.

To the east of the spacious sayr in which we had pitched our tents was a complex of exceedingly picturesque ravines in the so-called "blank" (unfossiliferous) series. These are bright-red sandstones which the wind and rain have cut through with deep, vertically sloping gullies and canyons. The Soviet paleontologists whose extensive excavations had preceded ours in the Nemegt had failed to find any fossils in this sandstone; they thus named it the "blank" series. Gradziński, on the other hand, who had spent three weeks at Nemegt with our team the year before in order to study the sedimentation conditions of the local sandstones, had concluded that the series was anything but "blank." He based his conclusion on the fact that it contained numerous accumulations of broken dinosaur-egg shells, which seemed to indicate that these were windborne deposits, i.e., that deposition had taken place on land and not under water. On the other hand, the somewhat lighter colored sands and sandstones overlying the Nemegt blank series and forming the slightly less steep ravine complex to the north and west of our camp were deposits that had been formed under water. Gradziński claimed that these contained dinosaur skeletons; as far as mammals in the Cretaceous deposits of Nemegt were concerned, they could be found nowhere but in the "blank" series.

The area we had to explore was immense. The Cretaceous sandstone ravines at the foot of the Nemegt range extend over something like twenty square miles. On July 31st, a day which was to go down in our history as our V-day, we had an early breakfast and went off in all directions. A small party went east to explore the unpromising blank series. Teresa departed for the north, to explore the light-colored sandstones near camp. Jerzy Małecki, who was fond of long walks, went north almost as far as the foot of the Nemegt Range. I myself went to the sayr group west of camp. On my first day in the area I had walked to the sayr farthest to the west and had found a few hind-limb bones and an ornithomimid pelvic fragment which certainly belonged to the same individual. I was now carrying a jug of glue in order to glue the bones together before their removal, which was scheduled for two days later.

We had been enjoying marvelous weather ever since our arrival at Nemegt. The rain that had bothered us at Altan Ula was no more, the sky was perfectly cloudless, and a cool north wind blew every morning; thus it was not too hot despite the sunny sky. The temperature never went above 80° in the shade, and we were able to work even at noon. When the north wind blew we would leave camp without water bottles, even for long trips, since we could then easily manage half a day without drinking.

Impregnating the ornithomimid bones with glue did not take much time, and when I was free to continue on my way it was still quite early. I walked along a shallow ravine among thorny caragana shrubs whose branches are covered in golden bark. The sun was still low and there was as yet no glare from the sandstone, whose light color made it difficult to tell bone from rock. At a certain point I spotted bones sticking out about 10 feet upon the left side of the ravine. On climbing the slope I saw a few small bones, no more than 10 inches long, protruding from a narrow sandstone ledge. Just beside the ledge, in a small crevice made by a seasonal stream, lay a sandstone block weighing several pounds; this contained randomly scattered fragments of narrow, broken ribs and two phalanges. The phalanges were only just over an inch long, so these had to be the remains of a small reptile. I turned the block over and gasped. The bottom of the block consisted of an almost complete skull about 10 inches long. The bones of the cranial roof were thick and rough, and the thick bones on the sides enclosed deep eyesockets. At first glance it looked like the skull of an armored reptile, but on closer inspection this proved to be a mistake. The thick, wrinkled bones on the cranial roof (which gave the impression of being chiseled) were not an external armor of plates or shields formed in the skin over the skull, but were simply thick, overgrown cranial bones. The sutures between the individual bones were clearly visible, and I could make out the outlines of the well-exposed frontal and parietal bones. The front of the skull had unfortunately been damaged, and the nasal bones were missing. The skull was relatively narrow and tall, and had two symmetrically spaced round holes at the back. Now, reptilian skulls bear so-called temporal openings, which are an important taxonomic feature and allow us to distinguish between skulls belonging to different orders. The Ruling Reptiles, which include among others the two dinosaur orders (Saurischia and Ornithischia), as a rule have two temporal openings, one upper

Ravines in the so-called blank series . . . 153 Nemegt, 1965

and one lower, on each side of the skull. This type of skull structure is called diapsid. The shape and size of the temporal openings vary in the different dinosaur groups. In the armored ones, which my reptile seemed at first glance to resemble most closely, the upper temporal opening is always overgrown. I turned the skull over and began to look at it from the side, but the lower jaw was unfortunately missing. On one side I could see a part of the upper jaw, with a row of tiny, fanlike teeth. Teeth of this type with vertical or fanshaped grooves, appear in various groups of order Ornithischia. The lower temporal openings were hidden on both sides by rock. Their shape would be seen only after the skull was cleaned in the laboratory. Nevertheless, it was clear that my small dinosaur was an ornithischian. The saurischian dinosaurs include, as we know, the herbivorous sauropods and the carnivorous theropods. My reptile was neither a large herbivorous sauropod with a small head on a long neck, nor on the other hand a large carnivore. Now, the other dinosaur order, the Ornithischia, is far more differentiated. Their heyday was the Late Cretaceous, during which period they were represented by the armored dinosaurs, the horned dinosaurs, and the highly differentiated group called ornithopods. The ornithopods included a number of bipedal, herbivorous, medium-sized reptiles of greatly specialized internal and external structure. Since what I had found was neither an armored nor a horned dinosaur, by process of elimination it had to be an ornithopod. I tried to recall the skull structures of the ornithopods with which I was familiar, but the skull wouldn't fit into any of the families I knew. At last I had the right idea. An ornithopod family called Pachycephalosauridae, which had a very strange skull structure, is known from Upper Cretaceous deposits in North America. In this group the bones of the cranial roof can be up to 10 inches thick. Pachycephalosaurs had a highly arched cranial dome above the forehead, probably for protection against their carnivorous enemies. While my skeleton was definitely not that of a pachycephalosaur, there was a certain resemblance. I recalled that one particular pachycephalosaur (of genus *Stegoceras*) had similarly placed upper temporal openings in the back of its skull, behind the thick, arching dome. In order to determine the skeleton's parentage we would have to prepare the skull, i.e., separate it from the enclosing rock, do likewise with the remaining available bones of the skeleton, and study the hundred or so published papers on ornithopod dinosaurs. (Subsequently, on our return to Warsaw, we did in fact establish that the skeleton represented a new genus, which might belong to Pachycephalosauridae.)

I was so preoccupied with inspecting the skull and mulling over its relationship to the various groups, that I failed to notice a viper sunning itself on the very spot in which I intended to sit down to clean the bones protruding from the block. It was distinctly disturbed by my legs and was moving about uneasily, shooting out its forked tongue again and again. Luckily enough, I had thick boots on: the last thing I wanted to do was to fight the viper or to kill it. Still, it was in my way, and the crevice in which I was standing was decidedly too narrow for both of us. My colleagues at the Institute for Ecology of the Polish Academy of Sciences in Warsaw, who catch vipers in the Kampinos Forest, had once shown me that if a viper is held up by the end of its tail it is quite defenseless, since it is incapable of raising itself up to bite one's hand. I was quite prepared to take the ecologists' word for it, but much less ready to grab the viper by the tail. Verbal persuasion and requests that the viper go away proved equally fruitless. It also failed to react when I threw pebbles at it, or at least its reaction was not what I intended to produce. This left me no choice but to push the viper down the cliff with my hammer. I did it clumsily, however, loosing a few stones at the same time, and one of these struck the viper. I jumped down to have a look at my victim. Its skull had been crushed—it could not even be taken back for preservation as a museum specimen. Having thus unintentionally killed one reptile, I proceeded, with a bad conscience, to bring the other back to life. I impregnated the skull with the remaining glue and put it in a safe place, waiting for the glue to set. Now I could at last sit down in the narrow, uncomfortable crevice just vacated by the viper and proceed to uncover the rest of the skeleton. Besides the hammer and a few chisels, I had also brought along a spatula and a small brush, and these proved exceedingly useful. I had to work in a very uncomfortable position. Another part of the skeleton was resting on a cliff ledge about 16 inches wide. The S-shaped spinal column, a part of which was soon exposed, was nearly 4 feet long (the whole animal was probably about 7 feet long). Removing the sand with the spatula and brush, I gradually exposed first the pelvis and then the small femur and the two parallel shin bones; the forepart of the skeleton was very incomplete.

After a few hours' work a large part of the skeleton lay exposed, and by 1 P.M. I set out on my return journey. As I neared camp I noticed that everyone had come back and that they were all sitting together at the tables under the awnings, carefully inspecting certain objects. These were some very fine teeth and jaw fragments, which the group exploring the blank series had found and which were now being impregnated with polystyrene. I sat down at a table and produced a magnifying glass. The objects were small lumps of red sandstone, very much like the red concretions from Bayn Dzak in which we had found skulls of primitive Cretaceous mammals and lizards. We carefully inspected the jaws, which were about three-eighths to five-eighths of an inch long. Some were very

Excavation of a large tarbosaur in Nemegt. Skull is visible in foreground, pelvis in background. Sitting, in front row: T. Maryańska and W. Siciński; background, left to right—Z. Kielan-Jaworowska, Gunzhid and Lam Dzal

Tail and pelvis of a large tarbosaur in Nemegt

well preserved, but all had tiny teeth, all of equal size, and were thus small lizards and not mammals. Finding lizards in the blank series was a valuable discovery, but finding mammals would have been even better. During our several weeks at Nemegt we explored the blank series repeatedly, but no mammals were ever found. Nevertheless we remained convinced they must be there and that they will be found by some future expedition.

We were all so impressed at finding lizards in the Nemegt blank series that we talked of nothing else at dinner, and it was only after dessert that I could announce my finding the almost complete skeleton of a strange, small reptile. It then turned out that we had all been in luck that day. Teresa had found the pelvis and skull of a large carnivorous reptile in light-colored sandstones only five minutes' walk from camp (these proved on exposure to belong to the practically complete skeleton of a 40-foot-long tarbosaur). Near Teresa's skeleton Barsbold had found a small tarbosaur, while a complete ornithomimid pelvis with various single bones had been found near the camp. And finally, Jerzy Małecki had located bones belonging to a very large predator—but his skeleton was several miles from camp and would be hard to reach by truck.

On Sunday, August 1st, we were able to sit down at leisure to discuss our working schedule at Nemegt. There was not much time left. We were due to meet the western group in Dalan Dzadgad on August 20th, so our team's last group would have to leave Nemegt on August 17th at the latest. This meant that the camp would have to be dismantled and all fossils and equipment sent out before this date. Dovchin, who had gone to Ulan-Bator with Izh as soon as we got to Nemegt, in order to ask the Mongolian Academy of Sciences to help us out in transporting the collections, was due back on August 8th. The Mongolian laborers' contracts ran out on August 12th, which meant that those skeletons which involved the preparation of large monoliths would have to be extracted by that date. We were thus faced with the far-from-easy task of extracting all the skeletons found—the biggest of which was 40 feet long—inside of not quite two weeks. We had to lengthen our working day and abolish the Sunday rest. The extraction of my small dinosaur skeleton was easy. Teresa and I worked on it for two days, impregnating the loose sandstone matrix with thin polystyrene. By the time the glue had dried, the sandstone was very hard rock, providing excellent protection for the bones within; these blocks could be safely transported by truck.

On August 2nd, we proceeded to uncover the huge tarbosaur skeleton found by Teresa. I had hoped that Małecki's skeleton could be exposed at the same time, but this would require daily trips by truck, and the truck was not available. That day Rachtan had declared it had to be taken to Dalan Dzadgad to have its frame

Preparation of monolith

welded, since it had been threatening to split apart ever since Altan Ula. He had been measuring the crack every day and had come to the conclusion that it was no longer safe. Since Izh's truck had not yet returned from Ulan-Bator, we were left during the next few days with only the one Mongolian truck. This truck had to bring water from 15 miles away and carry plaster, boards, tools, and water to Teresa's skeleton; it was also due to make a trip to Tsagan Khushu to bring back a five-man party which had left a few days earlier with a tent, food, and a tank of water to look for small mammals in the Tsagan Khushu Paleocene outcrops. We were thus unable to conduct routine work far from camp, and decided to concentrate on Teresa's skeleton. We also extracted the incomplete skeleton of Barsbold's small tarbosaur, and the pelvis of the ornithomimid.

Teresa's skeleton was lying fairly close to the surface, resting on its right flank with its head thrown back and its tail slightly arched. The last 7 feet of tail were missing. The skull had been preserved in an abnormal position, with the lower jaw incomplete and pushed far forward. Half of the skull was disposed vertically, while the other half, lying under it, was horizontal. The two parts could not possibly be separated in situ undamaged. We had to make a crate large enough for both halves, fill up the empty spaces with plaster, and take back the skull as a single large monolith, which, when crated, proved to weigh about a ton.

All that week the wind kept blowing in strong gusts, making the excavation more difficult. It was a particular nuisance in preparing the monolith. Where we used to grumble about sand in the eyes, which caused inflammation or even conjunctivitis, we now found that sand in the eyes is child's play compared to plaster in the eyes. If we interrupted work on that account, however, we would have to leave the skeleton behind. It soon turned out, however, that we would not in any case be able to take back the entire skeleton—not for the lack of time, but because parts of it were very poorly preserved. The cervical vertebrae, which lay at a shallow depth, were completely weathered, and disintegrated into powder. It would have taken at least 5 gallons of polystyrene to impregnate these bones, and even so the result would have been doubtful; moreover, we were running short of polystyrene. Unlike the neck, the other parts of the spinal column, along with the pelvis and the hind limbs, were in perfect condition. This was the largest tarbosaur we had ever found. The flat, elongated ilium was 4 feet long. On the abdominal side it articulated with the other two bones of the pelvis, the rather thin ischium pointing to the rear, and the pubic bone pointing obliquely forward. The tarbosaur pelvis, like that of all carnivorous reptiles, is triradiate, i.e., it consists of three bones—ilium, pubic bone, and ischium—all set at angles to each other. The pelvic girdle of the carnivorous dinosaur is very strong and has a characteristic shape. As in all bipedal reptiles, the left and right ilia in tarbosaurs were joined over the spine. Since tarbosaur skeletons are very common in the Gobi, we had learned to identify them even from only a small exposed fragment of the pelvis.

The most characteristic part of the pelvic girdle in carnivorous reptiles is the pubic bone, which in a mounted skeleton points obliquely downward and forward and whose tip bears a very large protuberance perpendicular to the bone itself. In our skeleton this bone was more than a yard long, with the length of the thick, heavy lateral protuberance coming to about 16 inches. The pelvic bones accommodated the powerful legs: the left leg, which we had found at a shallow depth, was pointing backward, while the right one, which lay deeper, was strongly bent at the knee and shifted forward somewhat. The deeper-lying bones were embedded in hard sandstone partly formed into conglomerate, from which they had to be extracted with large chisels and hammers. Four persons worked on the pelvis and hind limbs of the tarbosaur, two on the head, and two on the vertebrae of the tail. At the same time two laborers were digging around the skeleton in order to remove the overlying layers of sand and sandstone. We were quite crowded and in each other's way, but the need for haste made this unavoidable. We worked twelve hours a day and sometimes even longer. In a week's time the skeleton was almost totally exposed and polystyrene-reinforced, and so could be drawn to scale; after this the bones could be numbered and removed one by one. The main problem was how to handle the pelvis. We had separated the right and left ischia and the two pubic bones from the ilia. In the process the bones had split into several parts, which we had marked and crated. The main difficulty was how to pack the ilia. These were very large, thin, flat bones with a row of sacral vertebrae in between. Dorsally they almost touched, but abdominally they were about 28 inches apart. We did not want to risk separating these bones from the rock, for fear they would crack and be difficult to reconstitute in the laboratory. We therefore had to take them together with the rock, as a monolith. We took the exact measure of the crate required and made a suitable frame. This frame, 28 inches tall and 4 feet long, was put over the bones in the rock. Before starting on the preparation of the monolith, we assembled a few sacks of plaster, a barrel of water, buckets, and lignin around the skeleton. We covered the bones with plastic foil and began mixing and pouring the plaster. This had to be done in a hurry, so as not to allow one plaster layer to set before the next was poured; otherwise the monolith would crack along the contact surfaces of the layers during transport. To keep the plaster from leaking out, we padded the base of the frame with a dense layer of sand, but even so it absorbed plaster like a bottomless pit and looked as though it would never fill up. We poured in one bucketful after another, but the white surface rose only very slowly. To reduce the monolith's weight, we put in pieces of wood, dry *Haloxylon* branches, and even empty food cans. Finally, after we had put in several hours' work and 700 pounds of plaster, the frame was filled. We let it stand overnight for the plaster to set, and the next day nailed a stout wooden lid onto the frame enclosing the white mass; the monolith was practically finished. We now had to turn it upside down, fill the voids with plaster, and nail down the other lid. We dug a trench around the base of the crate, passed ropes around the bottom corners, and began turning the crate over. Eight persons pulled on the ropes from one side, while a few more pushed the crate with all their might. The monolith shuddered, the rock enclosing the bones broke off its base, and the crate was now free to be turned over. This operation had not always been successful, and we had had occasions on which the bones had broken in half, leaving pieces stuck in the base rock; but this time the monolith was very satisfactory. We filled the voids with plaster, nailed down the lid, wrote "This End Up," and marked the block on all sides with large numerals; the one-ton crate was now ready to go.

The tyrannosaur skeleton took up six very large crates which together weighed about 4 tons. Fortunately, the truck was able to drive up to the skeleton, for we had built about 25 yards of road from the skeleton to a stone-reinforced ramp we had dug in the lower part of the slope. Once backed up against the ramp, the truck could be loaded.

On August 8th Dovchin returned from Ulan-Bator, bringing good news: the Mongolian Academy of Sciences had promised to send two 8-ton Zil trucks with winches to Dalan Dzadgad on August 20th to carry the specimens to Ulan-Bator. We were all greatly relieved; the problem of getting our very large collection of crates from Dalan Dzadgad to Ulan-Bator was now solved. All we had to do now was to finish the work in Nemegt as quickly as possible, so as to have all the collections in Dalan Dzadgad by August 20th.

By August 10th most of Teresa's skeleton was already crated and only two monoliths were still waiting for the plaster to set. We could at last begin extraction of the skeleton found by Jerzy Małecki. A large group of us set out on foot, taking picks and shovels, hammers, chisels, and polystyrene.

The weather was beautiful, with a strong north wind and an absolutely cloudless sky. After we had been walking in a north-westerly direction for an hour we started grumbling, and asked Jerzy how much longer it would be—after all, he had said it was only a 45-minute walk. "I had to say that or you wouldn't have come," he replied calmly, "as a matter of fact, it's another two hours." Fortunately for him, he was only joking, and we were actually nearing the find. There were the large bones sticking out two feet down on the light-colored sandstone wall of a small hill. On the other side of the hill there were loose blocks containing other bones, and in another outcrop some 20 yards away were dinosaur tail vertebrae.

The fact that the bones were in three separate placed indicated that they could not all have belonged to the same individual, but all lay in the same bone-bearing level. We got to work, and soon had a dozen or so tail vertebrae of a carnivorous dinosaur.

The greatest interest seemed to lie in the large dinosaur bones we had seen first. Even though they were a mere two feet below the surface of the hill, they were not easy to extract. The top of the hill was a layer of strongly cemented sandstone, forming a slab about a foot thick. The bones showing on the surface of the wall could be extracted quite easily, but the rest of the skeleton was inside the hill and under that layer.

That day we made a special effort to extract as many bones as possible. We found that the bones in the wall were the foot and the metatarsus of a large tarbosaur. The leg was strongly drawn up; after removal of the metatarsal bones we saw the two parallel shank bones inside. Nearby in the same wall we discovered a few vertebrae from the same skeleton. We could not tell what else lay inside, since our picks could not manage the sandstone slab; this meant we had to come back later with heavy crowbars.

Not much time was left for extracting "Małecki's leg," so we decided to return to the spot next day by truck, taking a roundabout route and bringing a large supply of tools and packing material in order to try to extract as much of the skeleton as possible. We already knew that even if we should find an almost complete skeleton we could not manage to extract it, since the Mongolian laborers were due to leave Nemegt on August 13th and the camp had to be wound up immediately afterward so as to keep to the termination dates which had been laid down.

On the morning of August 11th everyone in camp climbed onto the truck and we set out on a 5-mile roundabout route to the skeleton. But this was not our lucky day. About a mile from the skeleton the truck broke down. The front wheel started screeching, there was a series of loud noises, and the truck stopped. The trouble turned out to be serious. We collected our equipment and continued on foot, leaving Aleksander Nowiński and Wojciech Skarżyński behind to help Rachtan with the truck. This greatly reduced the group that was free to work on the skeleton. We would now be unable to take the crates to the skeleton and to crate the bones in situ, as had been planned; the extracted bones would now have to be carried back to the trucks on our backs or on stretchers. We split up into three groups: one to extract the skeleton by laboriously breaking up the slab layer, a second to pack and number the bones, and a third (consisting of three persons) to keep moving between the skeleton and the truck with the packed bones. By nightfall, about 9 P.M., we had one perfectly preserved tarbosaur hind limb, a huge pubic bone, a large part of the pelvic bone, fragments of ischium, and about a dozen vertebrae. The remainder of the skeleton—we did not know how large a remainder—lay still imprisoned in the hill and had to be given up. When we returned to the truck at 9:30, Rachtan was just replacing the repaired wheel; we could now go back to camp. That night we went to bed at 11 P.M., after 15 hours of work.

The following day was also a busy one. We spent the morning packing and loading the last monoliths of Teresa's skeleton, and at 2 P.M. invited our Mongolian colleagues, who were leaving the following day, to a farewell party. We had become very friendly with them during our two months of sharing daily work and daily worry; they were excellent workers and we were sorry to see them go. We took parting photographs, exchanged gifts and addresses, and promised to write.

On August 13th the Mongolian truck left camp with Dovchin, Namsray, the driver Dzhamba, and all the laborers. We made an appointment to meet Dovchin in Dalan Dzadgad in six days' time. Barsbold was the only Mongolian left, and he moved to our camp for the remaining few days.

Our own group was also depleted. There was still a lot to do in Dalan Dzadgad, where part of the sauropod skeleton had arrived uncrated from Altan Ula in its enclosing rock. As the crating had to be done before shipment to Ulan-Bator, Mr. Rachtan and his truck left for Dalan Dzadgad, taking Teresa Maryańska, Wojciech Skarżyński, Aleksander Nowiński, and Łepkowski. In addition, Jerzy Małecki and the driver Izh left for Dalan Dzadgad with a collection of specimens. This left only five people in camp: the two geologists Ryszard Gradziński and Jerzy Lefeld, plus Wojciech Siciński, Barsbold, and myself. We still had much to do. The geologists had to check certain measurements and prepare the final sketches, the remaining specimens had to be numbered and packed, and the camp had to be dismantled. They were melancholy, those last few days; we knew this was the end of the expedition. Our valley, until now cheerfully noisy, became still and deserted. The days, which had seemed long at the beginning of the expedition, had passed so quickly that we were surprised to find our eleven weeks in the Nemegt Valley were over.

On August 17th two of our trucks returned from Dalan Dzadgad; by the 18th the camp was dismantled, the trucks were loaded, and our last group left the hospitable Nemegt Valley.

On our arrival in Dalan Dzadgad two days later we found that the four persons of the advance party had done an excellent job. The loose sandstone blocks containing the sauropod bones had been packed in 15 huge crates, with the empty spaces filled with plaster. The entire sauropod skeleton thus occupied 35 crates, the largest of which weighed more than a ton. Our specimen depot in Dalan Dzadgad was truly impressive. Even though a large part of the collection had already been sent to Ulan-Bator, we still had about 100 crates, weighing over 20 tons in all.

15. The Rhinoceroses of Altan Teli

Yak

The greatest surprise that awaited me in Dalan Dzadgad was a letter from the western group. Ever since we had parted in Altan Ula on July 12th, we had had only two letters from Halszka Osmólska: the first, sent from Dalan Dzadgad, had reported that a serious breakdown of the truck had forced her to stop over for a few days at Dalan; the other one was from a stop in Chandman, on the way to the Miocene outcrops at Begger Noor. It had been very cold in Chandman, which is over 10,000 feet up in the Bayan Tsagan mountain range. They had passed the night in a yurt hotel and in the morning had found the area covered with hoarfrost.

The letter I now received, dated August 5th, 1965, had been mailed in Altan Teli. Halszka wrote as follows:

"We left Chandman on July 21st and arrived in the valley of Lake Begger Noor. It was hot again; the countryside was quite green, and in the somon of Begger Noor we even saw a small potato field. The Darga of the somon gave us a very cordial welcome. We learned from Dashzeveg, who acted as interpreter, that the Darga knew of outcrops in the vicinity containing, as he put it, 'petrified bones.' So we set out with a guide and set up camp in a beautiful spot on the bank of a stream, quite close to a lone yurt and to a herd of grazing cows. The countryside was unusually beautiful and green. I don't know if I should be writing you this, because if you get this letter in Nemegt you'll turn green with envy, but we dammed the Begger Noor stream and were rewarded with a pretty pool the size of a large bathtub, with the water coming up to our knees. The place was charming, though it had one major defect: for the first few days we couldn't find any fossils worth mentioning. It was only on the fourth day, after we had thoroughly explored the whole neighborhood, that we found some well-preserved mastodon and rhinoceros teeth on the other side of the valley, in outcrops of Miocene gravel and loosely cemented sandstone. We therefore moved camp to these outcrops and began collecting sandstone concretions, which, when broken, often yielded bone fragments. But there was little paleontological material, and the bone remnants were usually in bad condition, so we decided to apply for permission to enter the quarantined areas and to reach the Pliocene deposits at Altan Teli.

"On July 26th Kuczyński and Walknowski made a two-day trip to Yessen Bulak and came back with the permission of the aymak authorities. After consulting with our Mongolian colleagues, we decided to pack up at Begger Noor and leave for Altan Teli as quickly as possible. Dashzeveg, like ourselves, was dreaming of rich bone beds. The day we left we were visited by two Soviet geologists doing stratigraphic work in the area. They knew the Altan Teli outcrop and gave us valuable advice on what route to take and on which side to start looking. Late in the afternoon on July 29th we left for Altan Teli. The further west we went, the greener the countryside became. We often met stockbreeding teams in the wide valleys, but there aren't many camels here; the Mongolian cowherds in the west mainly breed cows and yaks. The nomads with whom we spoke told us that most of the cattle bred in these areas were crossbreeds of cows with yaks. The animals didn't look much like cows, however; they were much shorter and stockier, were covered with long black or gray hair (often speckled), and had long horselike tails reaching almost to the ground. There were streams flowing through some of the valleys, and the grass was a luscious green, with the meadows covered with dark-blue flowers similar to the gentian in the mountains back home.

"We got to Altan Teli on August 1st, i.e., five days ago. The Altan Teli hills run along the northern fringe of the Dzereg Valley, at the foot of the Ömnö Khayrkhan range. On the other side of the valley we can see the austere, bare range of the Mongolian Altai, covered with white patches of snow. It is a most beautiful, majestic view, quite different from what we were used to in the southern Gobi. The nights are already very cold and we wear padded suits in the evenings, but the days continue to be very hot. However, it is neither the cold of night nor the temperature differences that we mind most; our main enemies are the swarms of mosquitoes which come from the streams running in some of the Altan Teli sayrs.

"We found the bone deposit almost at once the day after our arrival. The Soviet paleontologists weren't exaggerating when they described the outcrop as a veritable cemetery of Pliocene mammals. The bone layer is about 5 feet thick; we investigated it for over a mile and it seems to be equally rich everywhere. It would also seem that Rozhdestvensky was right in thinking that the thousands of animals buried here were killed by a flood. The five-foot-thick layer consists almost entirely of bones. We have been finding numerous complete Pliocene rhinoceros skulls, which Henryk Kubiak identified as belonging to the genus *Chilotherium*. Some are over two feet long, and there are also smaller skulls from young individuals. Hipparion and gazelle lower jaws and skulls are somewhat less common. We are also finding numerous long bones and other skeletal fragments, but no complete skeletons, since the bones are too crowded and intermixed. Our problem is in separating the skulls from each other, since there are so many that to extract one undamaged we have to damage several others. We are making monoliths for the larger rhinoceros skulls, and saturating the other bones with glue and packing them in lignin and gauze.

"We don't have much time. I understand from your last letter that we are to meet in Dalan Dzadgad on August 20th; besides, we have to get our own specimens to Ulan-Bator before that date

Excavation area at Altan Teli

169 Rhinoceroses of Altan Teli

Dalan Dzadgad. Loading collections onto trucks. All the members of the expedition worked like stevedores

On our way back we spent the night in
the ruins of a lamasery

if we are to arrive in Dalan Dzadgad with an empty truck and help carry some of your team's material. We are therefore going to terminate work at Altan Teli on August 15th, take our collections (about 20 cases) to Ulan-Bator, and then come to Dalan Dzadgad. We hired two laborers in Dzereg to remove the thin weathered layer above the bones; this is a great help, since it leaves us free to concentrate on selecting, preserving, and packing the bones. Yesterday, Kaźmierczak found a thin layer of turtle shells in a crevice nearby but there don't seem to be many of them.

"Altan Teli is truly beautiful, and it has so many perfectly preserved bones that we would be completely happy if it weren't for these frightful mosquitoes; they are making life miserable for us. The worst sufferer is Erdenibulgan, who is swollen all over. In the evening we light our so-called "argolators" (a device Walknowski invented, consisting of a can filled with glowing argol) and have our supper surrounded by clouds of foul-smelling smoke, but it doesn't help; the mosquitoes mind the smell much less than we do, arrive in droves, and get into our noses and mouths. We walk around with our faces wrapped up in gauze and smear ourselves with Taiga ointment, zip up our tents before going to bed, and hunt with a lighted candle for any mosquitos that have managed to get in. The veterans of last year's expedition claim that even the famous June flies at Naran Bulak were not quite as bad as the mosquitoes of Altan Teli.

"Walknowski is leaving today for Dzereg in order to bring a fresh supply of water and buy linen for wrapping up specimens—we have used up enormous amounts of linen and are beginning to run short of it—and he'll take this letter with him.

"If everything goes according to plan, we'll meet as planned in Dalan Dzadgad on August 20th."

We were overjoyed on receiving this letter. We had not been sure that our colleagues in the west would manage to reach Altan Teli, and we were also very glad to hear of the rich finds there. For Polish paleontologists, the Pliocene mammals of western Mongolia are particularly valuable. There is a spot in Poland, at Węże near the town of Zawiercie, with karstic caves in Jurassic limestones which contain a very rich bone breccia full of the remains of Pliocene mammals. The deposit was discovered and exploited just before the war by the late professor of geology Jan Samsonowicz. After the war a group of Polish paleontologists resumed work on the Węże breccia. Since the cementing material of the breccia was limestone, it was possible to extract the bones chemically, i.e., by treating the breccia with acetic acid; this does not destroy the calcium phosphate which is the main constituent of bones. By dint of several years of preparatory work carried out at the Museum of Earth Sciences and at the Paleozoological Institute of the Polish

Academy of Sciences in Warsaw, several tons of the breccia were dissolved, yielding a large number of well-preserved Pliocene mammals. Many papers on this fauna have been published by paleontologists working at the centers in Warsaw, Krakow, and Wrocław. There was thus a large group of experts on Pliocene mammals in Poland, and they would obviously be highly interested in the Pliocene fauna of Mongolia. Even though the Altan Teli deposit seemed to be older [Lower (?) Pliocene] than the Węże breccia, which corresponds to Upper (?) Pliocene, a comparison of the two faunas might nevertheless be expected to yield interesting results. The Altan Teli Pliocene fauna also captured the imagination of the Mongolian paleontologist Dashzeveg, who subsequently began working on a very interesting collection of the three-toed horses called hipparions drawn from these finds.

It was August 19th; according to Halszka's letter the western group was due in Dalan Dzadgad on the following day as planned. In the meantime those of us in Dalan Dzadgad were busy loading the crates holding the collections onto trucks. The trucks promised by the Mongolian Academy of Sciences had not yet arrived, so we were filling the four trucks on hand. This would have taken us several days, had we not managed to hire a hoist; with it, four 2-ton trucks could be loaded with crates and ready to go within half a day. Next day the two enormous trailer trucks arrived. Everybody in the expedition worked like a stevedore. Teresa and I kept circulating among the trucks, noting down the serial numbers of the crates loaded on the different trucks; the work went ahead very efficiently. The morning of August 21st arrived, but our colleagues from the west had not yet come. We therefore decided to accompany the crate consignment to Ulan-Bator, leaving two people behind in Dalan Dzadgad to await them.

As it turned out, that party had had a truck breakdown on the way and arrived in Dalan Dzadgad on the afternoon of the 21st, one day late.

We reached Ulan-Bator on August 23rd. The western party got there the same evening, but did not include everybody. The breakdown of their Star 66 had proved to be very serious, and Walknowski had refused to risk the exacting journey to Ulan-Bator. He had stayed behind in Dalan Dzadgad with Henryk Kubiak and the broken-down truck, and asked us to send a spare wheel to Dalan Dzadgad from our supply depot in Ulan-Bator. There is a regular air service between Ulan-Bator and Dalan Dzadgad, with two flights a week. We were in luck: the plane was due to leave the following day. We drove to the airport, and the pilot agreed to take the unescorted wheel in the passenger plane. Everything was set except for the weather. All of a sudden Ulan-Bator became swathed in darkness, as a raging sandstorm was unleashed upon the city. The plane was unable to take off, and a group of American tourists who were taking the plane to Dalan Dzadgad returned to the hotel covered with dust and disappointment. Next day the plane was at last able to depart, and Walknowski and Kubiak got their wheel in Dalan Dzadgad the same day. They returned to Ulan-Bator three days later, and we were all united once more.

Our return to civilization was quite amusing. We stayed at one of Ulan-Bator's best hotels, and some of our members, already dressed in city clothes, produced a general sensation with their beards, which they considered one of the major achievements of the expedition.

We still had two weeks of hard work before us. A committee of six, three Poles and three Mongolians, divided the collections between the Mongolian and Polish academies of science. The collections assigned to our side, as well as all the expedition's equipment and trucks, had to be packed and sent to Poland by rail. Our Mongolian colleagues helped us through all of this. On September 1st the equipment, the three trucks, and the collections belonging to the Polish side were finally sent off. Several dozen crates containing the collections assigned to the Mongolian Academy of Sciences remained in the yard of the Academy, shortly to be taken to the new premises of their Paleontological Laboratory in Ulan-Bator.

During the first few days of September we bade our Mongolian friends a cordial farewell and then (in two groups) flew back to Warsaw.

Members of the expedition, wearing street clothes after their return to Ulan-Bator. Left to right: E. Rachtan, W. Skarżyński, H. Kubiak and W. Siciński

Last farewell at Ulan-Bator airport. Left to right: Dovchin, Z. Kielan-Jaworowska and Dashzeveg

In the year that has elapsed since the termination of the field work of our last Gobi expedition, we in Warsaw have been preparing the skeletons of the dinosaurs collected, making plaster casts, and mounting those bones already free from their rock matrices. We estimate that it will take us about three more years to prepare all the material collected.

We are now faced with the most interesting part of our Gobi adventure: the scientific interpretation of our findings. But is it really the most interesting part? Was it not more exciting to travel through endless Mongolian steppes and extract skeletons of extinct animals in the burning sun of the Gobi than to make tedious anatomical studies, measure the bones, describe their shapes, look for nerve and vein openings in skulls, and pore over the abundant literature on these subjects?

No scientist familiar with the intellectual adventure of studying animals from times long past will have any hesitation in affirming that to travel millions of years back into the past, which is what paleontological study amounts to, is much more fascinating than the most exotic geographical travel we are able to undertake today. The study of animals that lived on Earth millions of years ago is not merely a study of their anatomy, but first and foremost a study of the course of evolution on earth and of the laws that govern it.

All the philosophical systems ever developed by man are essentially anthropocentric, with the tragic consequences for humanity which we know from history and which we are experiencing today. Studying the evolution of animals that inhabited earth for millions of years and comparing this with the history of mankind—so short on the geological scale—put man's position in the living world in its proper perspective and help counteract anthropocentric ideas.

The dinosaurs, which ruled on land for 155 million years, became extinct 70 million years ago. They are not the only group of animals to have done so. The group of marine arthropods called trilobites, which were the most common sea animals throughout the 350 million years of the Paleozoic era, died out 230 million years ago. Two large groups of marine cephalopods, the ammonites and the belemnites, died out 70 million years ago. Numerous highly differentiated mammal groups, including the multituberculates, the titanotheres, the dinocerates, the amblypods, and many others, died out during the Tertiary. These examples could be multiplied almost indefinitely. Many species have become extinct in historical times. If the history of an animal group is known, it is possible to tell which species are currently dying out, and will die out completely unless put under special protection. Extinction of species is just as common as their creation—these are merely two sides of the same process. Since the extinction of species and of large animal groups is a common evolutionary process, it is natural to inquire whether it might not in future affect those groups that are now at the peak of their development. Study of the reasons behind the extinction of species on our planet and of the laws governing this process may yet prove to be highly important to our own species in shaping its future development on earth.